I0492131

A Bunch of Sheep on Every Farm

by International Harvester Co.

with an introduction by Jackson Chambers

This work contains material that was originally published in 1918.

This publication is within the Public Domain.

*This edition is reprinted for educational purposes
and in accordance with all applicable Federal Laws.*

Introduction Copyright 2018 by Jackson Chambers

Self Reliance Books

Get more historic titles on animal and stock breeding, gardening and old
fashioned skills by visiting us at:

http://selfreliancebooks.blogspot.com/

Introduction

I am pleased to present yet another practical title on breeding and raising livestock.

The work is in the Public Domain and is re-printed here in accordance with Federal Laws.

As with all reprinted books of this age that are intended to perfectly reproduce the original edition, considerable pains and effort had to be undertaken to correct fading and sometimes outright damage to existing proofs of this title. At times, this task is quite monumental, requiring an almost total "rebuilding" of some pages from digital proofs of multiple copies. Despite this, imperfections still sometimes exist in the final proof and may detract from the visual appearance of the text.

I hope you enjoy reading this book as much as I enjoyed making it available to readers again.

Jackson Chambers

Introduction

NATIONS have taken up the implements of warfare and the wail goes up for food and clothing.

Hunger is being keenly felt by the peoples of Europe and the poor of our own land are in need of food.

All eyes are turned to America.

In meeting the emergency sheep occupy a prominent place.

No animal approaches the sheep in converting weeds and waste into wool and mutton. There is a wealth of food and raiment in the wasted grass and weeds of barn lots, fields, and roadsides.

Let there be a band of ewes with fat lambs on every suitable farm.

It is not the object of this booklet to describe alone the bright side of sheep raising. Our purpose has been to publish facts. The facts presented are taken from letters received from 5,000 farmers living in all parts of the United States, giving their experiences in sheep growing.

Textbooks and works on Live Stock Husbandry are valuable, but we believe that the actual experiences of men engaged in raising sheep, told in their own words, will be helpful to beginners as well as those of experience.

SHEEP ARE PROFITABLE

Reports From 5,000 Farmers Show Big Profits in Sheep Business

Out of over 5,000 letters received from practical sheep men from nearly every state in the Union, the Agricultural Extension Department of the International Harvester Company has compiled the following facts:

Of the farmers reporting, 3,750 live in Illinois, Ohio, Michigan, Pennsylvania, Iowa, Missouri, New York, Indiana, and West Virginia; 1,250 reports come from scattering states. Reports from those having range flocks in the west have not been considered.

Of the farmers reporting, 4,100 had farms of less than 200 acres.

Of the 5,000 farmers reporting, 4,250 had from 10 to 50 ewes; 4,000 farmers had ewes of the mutton breeds, the others had Rambouillets and Merinos.

Two thousand, seven hundred fifty farmers sold their lambs direct from the ewes without weaning them. The selling age was from three and one-half to five months.

One thousand, six hundred fifty farmers fattened and sold the lambs before they were one year old, or as soon as they were shorn of their first fleece. The others reporting either sold the lambs for feeders or matured them on the farm.

Two thousand, two hundred fifty farmers kept a few of the best ewe lambs each year for breeders.

Corn and oats were the grain feed for the ewes on practically all corn belt farms

Two thousand five hundred farmers bought wheat bran and oil meal to feed to ewes before lambing time and while suckling the lambs.

Merino Ewes and Lambs.   Owned by R. J. Henderson, Adena, Ohio

Clover hay and alfalfa hay was the choice of all for roughage for ewes. Many fed straw and fodder as a part of the roughage ration.

One thousand, two hundred fifty farmers fed silage to their sheep.

One hundred fifty farmers reported death of sheep from feeding mouldy silage.

An average of $4.69 a year was given as the cost for feeding a ewe, together with her lamb, until it was sold.

Each ewe returned an average income of $11.15, from the sale of the ewe's fleece and the sale of the lamb. (These figures were for 1916).

Three thousand farmers had lamb creeps and fed the lambs separate from the ewes.

Four thousand, twenty-five farmers did not feed the ewes any grain in summer.

Four thousand five hundred farmers advised having open sheds for the sheep except at lambing time.

One thousand five hundred farmers reported trouble or loss from stomach worms.

Where only a few sheep were kept and changed from one field to another no trouble from stomach worms was reported.

The remedies given for stomach worms were: 1—Change of Pasture. 2—Gasoline Treatment. 3—Worm Powders.

All but 18 of the 5,000 reported that "Dogs" were the main cause of the scarcity of sheep.

SHEEP RAISING A NEGLECTED PART OF AGRICULTURE IN UNITED STATES

Sheep in all countries of the world are decreasing in numbers.

There is a world-wide shortage of wool and mutton.

There is a growing demand for meat and wool.

Sheep have decreased in the United States 12,000,000 head in the past 17 years.

The range flocks of the west are rapidly being reduced.

Corn Belt farmers must help provide the future supply of wool and mutton.

It is patriotic—it is profitable.

SCARCITY OF SHEEP ALARMING

Ohio Sheep Buyer Unable To Make Purchases Of Breeding Stock From Western Ranges—Price Doubles But Will Go Still Higher

By George M. Wilber, Marysville, Ohio.

It is alarming to note how fast the flocks are decreasing in this country. Dairying in the agricultural sections, miners' dogs in the strictly grazing (hilly) sections east of the Mississippi River and the "nester" or homesteader in the west, have combined to drive out sheep raisers both on eastern farms and the western ranges until alarming conditions exist.

No Range Ewes For Sale

Last year I was not able to purchase a single car of breeding ewes west of the Mississippi River. All of the thousands of breeding ewes I handled were purchased east of Indiana. There are practically none for sale this year and the price has doubled in this section. Formerly I shipped in train loads from Oregon, Wyoming, Montana, New Mexico, etc., but it looks like there could be no ewes bought anywhere at a price which would attract the average farmer to invest; and yet even at the high price for ewes they would make more clear money than anything else. They are bound to increase in price even above the present values.

Small Flocks Fat and Free from Disease

There is positively not a farm east of the Mississippi River which could not profitably keep a small flock of ewes, 25 to 50 or more, which would not only trim up the weeds at no cost, but the owner would never know he had them so far as cost for feed is concerned. These small flocks are always fat and free from disease.

Neighborhoods Buy Car Lots

I would suggest that you urge farmers in their several neighborhoods to combine and buy a deck or car load and distribute among themselves. It is not practical to ship less than full cars, because car load rates are charged for less than car lots unless crated.

Sheep have decreased 50 million head in the world since the war began

---

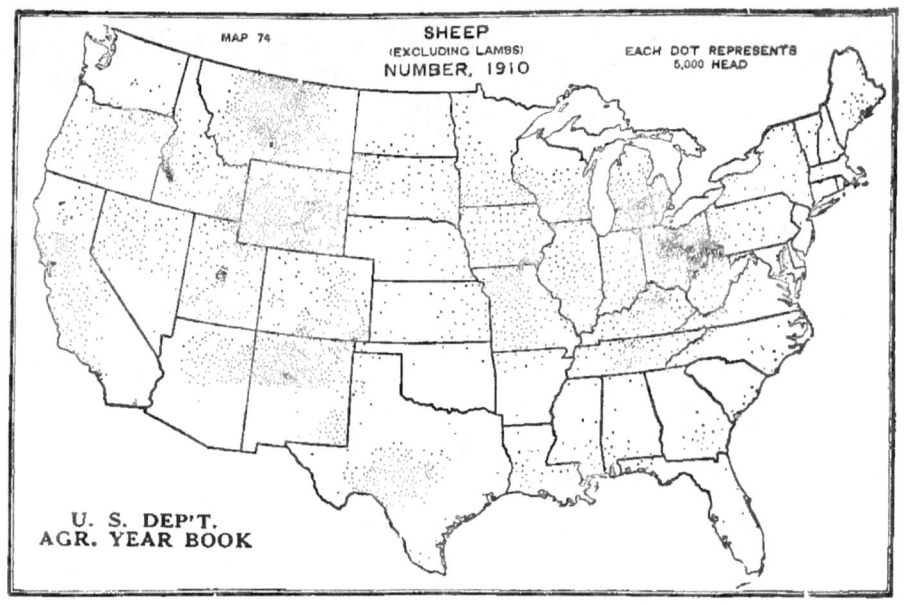

SHEEP IN UNITED STATES

State	Sheep Jan., 1917	Increase Since 1910	Decrease Since 1910
Alabama	121,000	....	22,000
Arizona	1,632,000	405,000	....
Arkansas	124,000	....	20,000
California	2,524,000	117,000	....
Colorado	1,950,000	524,000	....
Connecticut	18,000	....	4,000
Delaware	8,000	....	....
Florida	119,000	5,000	....
Georgia	150,000	....	38,000
Idaho	3,195,000	184,000	....
Illinois	898,000	....	162,000
Indiana	1,005,000	....	332,000
Iowa	1,240,000	94,000	....
Kansas	348,000	76,000	....
Kentucky	1,155,000	....	208,000
Louisiana	240,000	62,000	....
Maine	157,000	....	49,000
Maryland	223,000	....	14,000
Massachusetts	25,000	....	8,000
Michigan	1,834,000	....	472,000
Minnesota	541,000	....	97,000
Mississippi	193,000	....	2,000
Missouri	1,370,000	....	441,000
Montana	3,744,000	....	1,637,000
Nebraska	381,000	87,000	....
Nevada	1,455,000	300,000	....
New Hampshire	35,000	....	9,000
New Jersey	29,000	....	2,000
New Mexico	3,300,000	....	47,000
New York	840,000	....	90,000
North Carolina	140,000	....	74,000
North Dakota	250,000	....	43,000
Ohio	2,944,000	....	965,000
Oklahoma	104,000	62,000	....

Oregon	2,484,000	....	215,000
Pennsylvania	835,000	....	48,000
Rhode Island	5,000	....	2,000
South Carolina	30,000	....	8,000
South Dakota	658,000	....	47,000
Tennessee	650,000	....	145,000
Texas	2,328,000	519,000	....
Utah	2,089,000	262,000	....
Vermont	100,000	....	19,000
Virginia	686,000	....	119,000
Washington	585,000	109,000	....
West Virginia	715,000	....	195,000
Wisconsin	645,000	....	285,000
Wyoming	4,381,000	....	1,016,000

Total—48,483,000 in 1917 } Decrease—3,961,000
52,444,000 in 1910

Figures from "Agricultural Statistics."

SHEEP THE MOST PAYING ANIMAL

By George W. Grim, Fremont, Indiana

I think that sheep are the most paying animals that the farmer can have on the farm as they eat weeds and sprouts and keep down most foul weeds. A good bunch of ewes make money.

The sale of wool comes in the spring when most farmers are short of change, with a good bunch of lambs to sell later on and the old sheep are left. I would not be without a good bunch of ewes on my farm.

A BUNCH OF SHEEP ON EVERY FARM

1. They are Profitable.

2. They eat Weeds.

3. They Convert Waste Into Profit.

4. They Improve the Farm's Appearance.

5. They do not Require Expensive Shelter.

SHEEP PROFITABLE ON HIGH PRICED LAND

Practical Experiences of Nearly Three Thousand Farmers Prove Sheep Growing A Practical and Profitable Business in Corn Belt States—Converts Waste Products Into Profit—Cleans Up the Weeds—Expensive Shelter Not Required—Enriches the Soil

SHEEP RAISING PAYS IN IOWA

By E. L. Bitterman, Mason City, Iowa

Nearly every farmer in the corn belt can keep a flock of 25 to 50 head of sheep on a 160-acre farm with very small cost. Sheep can be pastured on oat fields and also many other fields in the spring when cattle are too heavy. We graze our oat fields every spring for ten days to two weeks, about the last of May before the oats are large enough to shoot. *After harvest they are run on the stubble where rape has been sown in the spring.* Thus our sheep are not on the real pastures only a few months in the year. I find many farmers like sheep. Sheep have their ups and downs like all classes of stock but on a high-priced Iowa farm our sheep have always paid as well as any other stock.

(Photo from American Shropshire Registry Assn., La Fayette, Ind.)

Sheep in Rye Pasture

PROFITABLE FLOCK FROM A SMALL START
By A. G. Marshall, Lancaster, Ohio

Purchase ten head of good open-wooled ewe lambs or yearlings, all as nearly of a size and type as possible. Have them all docked, put a bell on one of them. See that they are kept clean and well tagged, as no sheep will do well carrying a lot of filth. Turn these ten ewe lambs in the field with cows provided the field is free from cockleburs or burdock. Keep sheep out of foul barnyards or filthy stables to avoid foot rot.

With special care these ten ewes will bring $10 per head each year for lambs and wool. I started one man with three lambs—I sold them to him for $2. Dogs had half-killed them. In only a few years he had a bunch of 30 head and sold $100 worth of wool. Some of his sheep weighed 175 pounds. Every farmer should have a flock of sheep.

PROFIT OF $600 IN THREE YEARS
By J. B. Muchmore, Oblong, Illinois

I started three years ago with a small bunch and bought altogether 57 head in three years. I sell off my wool and male lambs and cull every year and I now have a bunch of 42 head of sheep and 30 lambs.

I consider I have for my feed and trouble a profit of nearly $600 for three years and my land has been improved by their use.

EVERY FARM SHOULD HAVE SHEEP
By Harlan Timmons, Morning Sun, Iowa

It would be profitable for every farmer to have a bunch of sheep. The manure is very rich and like hen manure, entirely free from weed seeds, and when the land is pastured, it is evenly distributed over the ground. Sheep are the most profitable stock I raise considering the money invested and the quick returns.

SHEEP NO EXPENSE ONE-HALF OF YEAR
By Frank McQuenn, Esmond, Illinois

I figure that my sheep cost me nothing six months of the year. *As soon as we are through threshing they are turned into the oat field and get no other feed until the snow comes.* I think my sheep are the best paying stock considering the good they do to the land. By all means give the boy good ewes to start with.

MADE 125 PER CENT ON CUT-OVER LAND

I Make More Money on Sheep Than on Anything I Have, Considering the Money Invested

By O. G. Puckett, Sauble, Michigan

I will tell you what I am doing here in the cut-over and burnt-over lands of Lake County, Michigan. I bought common ewes and have been using a Shropshire buck. I find that the coarse or medium-wooled sheep give me a greater percentage of lambs than the Merinos, but less wool. Last winter I fed five pounds of silage per day with bean hulls and a little hay. I fed no grain. I am using the sheep to help clear up the underbrush which has grown up here, which is a benefit to the land. For this reason I do not charge the sheep up with pasture.

There is more and easier money in sheep than in any other live stock, considering the amount invested. I made 125 per cent on a lamb crop this year. The lambs are not carried through the winter except the ewe lambs needed for building up the flock. When I have to quit raising sheep I will quit the farm. A good many of my neighbors want sheep but they cannot find any here for sale.

SHEEP GREATEST PROFIT PRODUCERS

By Wright & Wright, Bridgewater, Virginia

We keep a rather decent set of books and can show you that the sheep have given the highest per cent of profit of any stock kept on the farm. This has been true for seven or eight years.

Shropshire Lambs on Rape Pasture at Purdue Farm,
LaFayette, Indiana

STARTED IN SHEEP BUSINESS ON $10
By Howard K. Keim, Ridgefield, Washington

A beginner bought five cull ewes for a ten dollar bill. The owner turned them in his horse lot and allowed them free access to an empty box stall containing a trough low enough to feed oats. These thin ewes were fed handfuls of oats daily all winter and had good clover hay and some grass. In early spring they dropped eight lambs, three of which were ewes. The best buck lamb was sold for $10 to a neighbor for breeding purposes. Four other lambs brought $17.50 from the butcher. The three ewe lambs were kept to increase the flock. The following spring before shearing and with seven small lambs at foot the lot was sold for $50 and the buyer made money on them. Names and dates could be given if deemed necessary. No other animals kept on the farm will return as large a profit on the investment as a well-cared-for flock of sheep.

SHEEP WILL PAY PROFIT OF 50 PER CENT
By A. A. Bates & Company, Irwin, Ohio

It's just as cheap and much more satisfactory to keep registered sheep as it doesn't cost any more in the end. Ten ewes and a ram at a cost of $275 will almost pay for themselves the first year. They will pay 50 per cent on investment for twenty years if properly handled and managed, but don't think because you have sheep you can expect them to give a good account of themselves unless you look after and feed them properly.

DO NOT BECOME DISCOURAGED
By John R. Nash, Tipton, Indiana

Start in the business when sheep are low in price, then you will not become discouraged. If you meet with a loss, just keep at it. All my life I have bred, fed, and showed sheep. No stock on the farm will pay as large a profit on the investment as a small flock of sheep.

A VIRGINIA REPORT
By R. M. Lawson, Burkee Garden, Virginia

Properly cared for they make more pounds salable food at higher prices than any other stock, food value rated. They should be kept on every suitable farm.

11

SHEEP PAY ON HIGH PRICED LAND

Reports from over 1,000 Corn Belt Farmers Show the Average Cost of Feeding a Ewe and Her Lamb to be $4.69 per Year—Lamb and Wool from Ewe Sell for $11.15

More than 1,000 corn belt farmers have given careful reports on the cost of feeding a good ewe for the year 1916 and her lamb to selling time.

The majority reporting sold their lambs when weaned. The average feed bill reported for ewe and lamb was $4.69. A few were very much higher and some decidedly lower than the average, as they credited the ewes and lambs with eating weeds and gleaning fields after harvest, but the great majority gave the cost very near the average figure.

Rambouillet Lambs

It is interesting to note that the figures furnished were from farmers who kept a small bunch of ewes that dropped lambs in early spring and the lambs were sold in early fall.

Not Guess Work

The reports were not guess-work. Many went into itemized detail of cost. The same men reported the gross income for the year from the sale of the lamb and the ewe's fleece. The average income was $11.15. Reports on income showed less variance than reports on cost. This fact verifies their reliability as in-

12

come is largely regulated by markets while the cost varies with farm conditions. One significant fact is that where the cost of feed was highest, there was almost without exception, a greater corresponding profit from sales of wool and mutton, showing that good feeding plays a big part in sheep profits.

The data on the cost of feeding the ewe and lamb, furnished by the farmers, is lower than that given in Experiment Station trials, due to the fact that the ewes and lambs in more than half the reports were not charged with the pasture in stubble fields, barn lots, weed patches, and roadsides.

The Pennsylvania Experiment Station, State College, Pennsylvania, gives some valuable information on cost of feeding breeding ewes in their bulletin on "Maintenance of Breeding Ewes."

Registered Sheep

Forty-five breeders of registered sheep reported an average yearly cost of $5.60 for ewe and lamb and a yearly income of $32.88 from the sale of the ewe's wool and the sale or value of her lamb.

PROFIT FROM SHEEP ON $200 LAND
By C. W. Bentley, Sauble, Michigan

I started with a pet lamb when I was three years old and have never been without sheep since and never expect to be as long as I live on a farm. If it had not been for my sheep I would not own a farm today. I expect to increase my flock to 200 breeding ewes as fast as I can.

I raised sheep on a farm in Ohio where land was worth from $100 to $200 per acre and I can truthfully say I made on an average of 100 per cent on money invested each year and have done even better since in Michigan. I might say further that I spread the manure from those sheep in Ohio on the meadows and the hay was 100 per cent better by so doing. I have found that sheep will live on most any kind of weeds but, like any other kind of stock, better feed, better sheep, better returns, and a great deal more satisfaction.

CAN KEEP TEN EWES WITHOUT COST

Eat Enough Weeds to Pay Their Way

By W. A. McDorman, Selma, Ohio

I believe every Corn Belt farm can keep ten ewes to each hundred acres without costing a cent. They will eat enough weeds to pay their way and the income will be clear gain. Sometimes we have to study where to put them! They can go in the meadow early in the spring and clean the white top from the hay. They can go in the woods lot a little later, then clean up the lots and yards around the barn and sometimes the roadside, then the stubble fields after threshing. This changing about will help to keep them healthy. If confined to one field or pasture they should have tobacco and salt kept where they can have free access at all times as a preventive of stomach worms.

SHEEP KEEP FAT ON RAG WEEDS

By S. S. Stettbacher, Alhambra, Illinois

I would advise having each field fenced, so that when the crop is harvested the sheep can be turned in and thrive in fields which otherwise would grow up in weeds. Good healthy sheep will thrive on rag weeds. When the corn in laid by the lambs are about ready to wean. *The cornfield is a splendid place for them if there is grass or weeds to be cleaned up such as morning glories and green cockleburs.* If the corn blows down they must be taken out when it begins to shoot.

(Photo from University Extension Service, St. Paul, Minn.)
Sheep Converting Brush and Leaves into Wool and Mutton

14

SHEEP CLEAN UP BRUSH LAND
By George Y. Tedrow, Guysville, Ohio

I would advise every farmer to try a small flock of ewes. They make the most money for the time and capital invested of any stock. I keep some stock cattle, some cows and hogs, and I think sheep pay best. They do not hurt the pastures nearly as much as some think they do. We could hardly keep this brush land in southeastern Ohio under control if it were not for the sheep. There are only a few kinds of weeds in this section that sheep will not eat. *I have had cattle pastures covered so thickly with iron weed, cockleburs, and briars, that the cattle could hardly be seen, and after pasturing with sheep a few years there were no iron weeds in the field and very few burs or briars.*

SHEEP EAT MORNING GLORIES
Advise Your Neighbor to Kill His Dog and Get a Bunch
By A. H. McKellar, Waterloo, Iowa

Give sheep a fresh pasture every year if possible. Sow rape in your oat fields. It will make a fresh pasture and also a very good one. Let them run in the corn fields. They will eat the leaves, grass, and morning glories, and clean everything up fine. Kill your own dog, and advise your neighbor to do the same. Some of the parasite enemies of sheep come directly from dogs.

KEEP DOWN NOXIOUS WEEDS
By Walter Casler, Ovid, Michigan

I always sell my lambs off of grass as soon as they are large enough to bring the top price. The sheep owner should have his farm fenced so that every field can be pastured. Not only will the pasture pay for the fencing in a short time but the sheep will keep down all noxious weeds, and change of pasture is one of the most essential things in sheep raising. I plan to buy a few new ewes every year or two, but if good ewes are kept it is a good plan to save the best ewe lambs.

SHEEP CLEAN UP CORN FIELD
By David Needham, Virginia, Illinois

I think it is profitable to keep a medium-sized flock of sheep on a farm where there is plenty of pasture, so the flock can be changed from one pasture to another. Use them to clean up stubble fields, pastures, and corn fields, where they will eat weeds and the lower blades of corn without damaging the grain to any extent.

15

25 EWES TO EVERY 100 ACRES
Convert Briars and Weeds Into Wool and Mutton
By G. D. Work, Galena, Ohio

I breed the pure-bred Delaine Merino—those big, smooth fellows that raise a handy-weight lamb. My wether lambs one year old in April, when sold in June weighed 80 pounds and brought $5.60. They sheared a 11-pound fleece worth $3.36. Merinos withstand the ravages of disease better, thus will stand closer herding. Of course, if I were wanting to raise a lamb to go to market at six months old I would prefer a mutton breed. If I were starting life over again I would start with a flock of sheep.

I convert all my briars and weeds into wool and mutton and sell it for first-class instead of selling it in hay as second-class material. If every farmer who owns 100 acres of land had 25 brood ewes to start on he would find he would make a larger per cent on money invested than anything he could have about him. My motto is, "Keep Sheep."

INDIANA FARMERS SHOULD HAVE SHEEP
By W. K. Franklin, Danville, Indiana

It might not be profitable nor practical for every farmer to maintain a flock of sheep, yet I think it safe to say that at least 90 per cent of the Indiana farms could profitably maintain a small flock.

(Photo from American Shropshire Registry Ass'n.)
Shropshire Lambs Being Fitted for Show. Note the Board Over Rack to Keep Them from Jumping Over Trough

16

BUYS LAMBS TO CLEAN UP WEEDS
By Wesley Brubaker, Ashland, Ohio

I formerly kept a flock of ewes the entire year, but for the last ten years have fed western lambs. I usually buy them in August or September and they clean up the briars and weeds on the farm for me, but this way of handling sheep is more risky and requires more capital than to get a start in a small way and grow.

I plan all my farm crops so as to have plenty of good feed to fatten lambs; nothing beats good clover, hay, and corn. I frequently sell small flocks of ewe lambs to my neighbors from my cross-bred Idaho feeding lambs and they always do very nicely. I have one neighbor who has fifty Delaine ewes from which he realizes every year $500 gross.

SHEEP IMPROVE LOOKS OF FARM
By C. R. Oder, Welton, Illinois

Keep the whole farm sheep tight, as sheep will clean out all weeds and brush that grow in the fence corners, and after the corn is eared out they can be turned in the field and will do no damage to the corn. Sheep will clean up a farm in less time than any other animal, excepting the goat. In driving along the road you can tell every farm which keeps sheep even if you don't see them. The fence corners are not full of brush and weeds and the lots are all free from weeds, all of which help the looks of any place.

CANNOT FARM WITHOUT SHEEP
By A. R. Jacob, Short Creek, West Virginia

I enjoy my flock of sheep, they keep the farm so clear of weeds. I think so much of them that I have often remarked, "When I have to quit keeping sheep I will quit farming." I give them full credit for what success I have made as a representative farmer as they are the only class of stock I can clip the coupons from and have the bonds left.

OPEN SHEDS IN TENNESSEE
By Perry Brown, Spring Hill, Tennessee

We have open sheds, and fields fenced so all can be pastured. A deep-milking, long-legged active ewe bred to a Dorset ram is the best for this country. Don't overstock.

17

SHEEP DO NOT REQUIRE EXPENSIVE SHELTER

More Than One Thousand Farmers Reported in Favor of Open Sheds For Sheep. Many Did Not House Them Except During Storms and At Lambing Time.

SHEEP NOT SHELTERED IN COLORADO

By Chas. I. Bray, Associate Professor of Animal Husbandry, State Agricultural College, Fort Collins, Colorado

Very few sheep men give shelter to their sheep except at lambing time. Alfalfa hay, silage and corn makes the best feed. We advise a man to start with 30 or 40 range ewes, one year old. Use a pure-bred ram and save the best ewe lambs. Buy two or three pure-bred ewes of the same breed as the ram and gradually work into pure-breds. There is a great need for breeders of pure-bred mutton sheep in this state.

DON'T KEEP IN TIGHT BUILDINGS

By A. R. Runyan, Rochester, Michigan

Don't keep sheep in a tight building. Better have an open shed. All they need is a roof and wind break. Don't compel them to eat musty or spoiled feed. Don't compel them to drink tainted water. Don't let them run to hay or straw stacks and get their wool full of chaff and then have to take less for it.

DO NOT REQUIRE EXPENSIVE SHELTER

By Poirson Bros., Fort Wayne, Indiana

We have no expensive shelter for our sheep. Just a well-drained open shed. They are given plenty of range and feed that we grow on the farm with a little oil meal and wheat bran that is purchased.

OPEN SHEEP SHEDS IN CANADA

By W. H. Beattie, Wilton Grove, Ontario

I have thirty sheep on 135 acres. I find that lambs dropped in March do best. I always keep my best ewe lambs and sell the rest when I get a buyer. Have open sheds facing the south. Feed alfalfa, hay, and roots. A man that does not like sheep is better without them. No stock pays better than sheep. They eat all kinds of weeds and keep the land rich.

18

$1,100 PROFIT IN FOUR YEARS

Experiences of a Texas Farmer Who Grew into the Business
By W. H. Ransberger, Coleman, Texas

Seven years ago I bought 23 head of Merino sheep and put them on my 320-acre farm. I had a 20-acre pasture fenced sheep tight. The fourth year my flock clipped 1,215 pounds of wool that sold for $216.50. As sheep increased, I fenced more pasture. At the end of the fourth year I sold out at a net profit of $1,100. I then purchased six ewes and a ram, pure-bred Hampshires, and made a net profit of $107 the first year. Following my plan of growing into the business I think any farmer can be successful with sheep.

PROTECT FROM RAIN AND SNOW
By W. George Cavan, Sugar Grove, Illinois

Have sufficient shelter to protect them from wet and snow. That is all. Select the breed you like best; have a good ram; you will scarcely miss what sheep eat. Always keep your best ewe lambs unless you can buy better ones elsewhere.

(Photo from Farmers Advocate.)

A Pole Shed Covered with Straw Makes a Good Shelter. Straw Must Be Stacked so as to Turn Water. Sheds Must Be Built on a Well-Drained Location

VENTILATION IMPORTANT
By J. M. Walker, Middletown, Ohio

I have kept sheep five years and they have paid me larger profit on money invested than any stock on my farm. The trouble in this rich farming country is that the average farmer thinks they are too small to bother with. Give them a well ventilated shed for shelter.

OPEN SHEDS AND CHANGE PASTURE
By C. M. Elkins, Prineville, Oregon

We have 200 sheep on 320 acres. Have open sheds and change pasture frequently. Sheep like some weeds better than they do grass. Sow rye in September for early spring pasture. If a sheep bloats give one pint of milk fresh from cow. If milk is not fresh, put in a teaspoonful of turpentine.

ORDINARY SHEDS
By Robert F. Miller, University of California, Berkeley, California

Have ordinary sheds for this climate and fence the farm so as to turn into any field. Start with good grade ewes. Don't buy overfat ewes. They may be barren. Sell off the broken-mouthed ewes and those with spoiled udders.

Ewes Cleaning up Grass and Weeds on Roadside

20

SHELTER AT LAMBING TIME
By W. E. Green, Francisco, Alabama

They need good shelter at lambing time. Sheep are scarce in this country. Give them a change of pasture and good feed and they destroy harmful weeds and bring a nice income. Last August I bought five common ewes and one ram for $21.25. I raised seven lambs and sold them and the ram for $51 and wool for $10.50. I teach the lambs to eat meal when they are young. There is more profit in sheep than in cattle or hogs.

HAVE A WELL-VENTILATED SHED
By Fred E. Reichert, Ann Arbor, Michigan

I would have farm fenced so the sheep could have the run of different fields to avoid stomach worms. Keep them in a well-ventilated shed or barn. I was brought up with sheep and have yet to see the time when they would not pay for the little extra care given by a good herder.

HAVE A WINDBREAK FOR SHELTER
By W. H. Edick, Pray, Montana

A good windbreak for shelter but not a warm shed. Better let them lie in the snow than in a close shed where they will steam and get catarrh. Give them a dry place to lie. They require more attention than horses or cattle but they hand you the pay oftener.

NEVER TURN SHEEP OUT IN RAIN
By Jacob Goebel, Charlestown, Indiana

I always keep my best ewe lambs. Change pastures every two or three weeks—never turn sheep out in rain in winter. I am 71 years old and have raised sheep for over forty years and have made more out of sheep than any other stock. From my sixteen sheep I sold last year $169 worth of lambs and wool.

(Photo from Farmers Advocate.)
Good Windbreak for the Flock

21

DON'T USE A SCRUB RAM

Lambs Sired by Good Ram Sold For $7.35 Per 100 Lbs. Those Sired by a Scrub Ram Sold For $4.50

In 1913 the Missouri Agricultural Experiment Station conducted a test to show the value of a good ram.

Thirty-four Colorado ewes were selected. They were uniform in size and condition and showed a predominance of Merino blood.

Grade Western Ewe   Good Hampshire Ram

(Photos from Missouri Experiment Station.)

Lamb from Hampshire Ram   Lambs Like this Sold for $7.35 per 100 pounds

Scrub vs. Registered Ram

Seventeen were bred to an inferior "scrub" ram and seventeen bred to a fairly good registered Hampshire ram. The lambs were sold at three months of age. The lambs sired by the Hampshire ram made 26.6 per cent more daily than those sired by the scrub ram.

It took 52.81 pounds of grain to make 100 pounds of gain on

the lambs sired by the Hampshire and 88.78 pounds of grain to make 100 pounds of gain on the lambs sired by the scrub ram.

Grade Western Ewe

Scrub Ram

Difference in Lambs

The lambs were sold at East St. Louis stock yards. The Hampshire lambs sold for $7.35 per 100 pounds, and the scrub lambs sold for $4.50 per 100 pounds. The well-bred lambs were thicker-fleshed, smoother, broader-in-back, and tighter-in-pelt than the scrubs.

A few dollars extra spent for a good ram means many dollars in-creased value in the lambs.

Lamb From
Scrub Ram

Lambs Like This Sold for $4.50
per 100 pounds

Good Ram More Than Half the Flock

Select a ram with a short, thick neck, big, masculine head, wide between the eyes, a prominent poll or crown and a short, broad, clubby nose.

A long slim-nosed ram with thin neck and low head will prove a disappointment. Be sure to have a ram with good mutton form, straight back, round-ribbed, full hind quarters and wide, deep chest.

Watch his eating and way of moving. One that chews rapidly and is alert and quick of motion is vigorous and usually a good stock ram.

GETTING A START WITH SHEEP

Don't Start With Too Many—Keep One Breed—Cull the Ewes—Save Best Ewe Lambs

Better start with a few ewes and grow into the business. With good care sheep increase rapidly.

In a few years a good-sized bunch of ewes can be grown from a small start.

Pure-Bred Ewes

Sheep need not be pure-bred or registered to be profitable. The average farmer will do better producing sheep and lambs to sell for mutton than producing pure-bred sheep to sell for breeding stock.

Selling Breeding Stock

The farmer who is a good judge of sheep, understands breeding pure-breds, and is a good salesman, can make the breeding of pure-bred sheep very profitable.

Have Ewes All One Breed

Whether the ewes are pure-bred or not it is best to have them all of one breed. They look better, their fleeces will be more alike, and their lambs more uniform than from a mixed bunch. The breed is a matter of choice. All of the well established breeds have good qualities.

Kind of Ewes—Sort out Delicate Ones

Discard ewes with small bodies, narrow chests, and those that are small around the heart; also the ones with crooked and club feet that are apt to catch and hold filth between the toes.

Ideal Type of Breeding Ewe. Note Short Neck, Straight Back and Roomy Middle. Property of J. C. Andrew, West Point, Ind.

Keep Big Blocky Ewes—Save Best Ewe Lambs

Sheep that have long, thin, "goose" necks usually have weak constitutions and are poor feeders. Select up-headed, stylish ewes with broad backs, roomy bodies and plump hind quarters. Have them uniform in size, mate them with good rams of the same breed, keep the choicest ewe lambs, and the farm flock will improve in value and returns each year.

Young Ewes

There are several ways of getting started with sheep. One way to start is to buy a few good young ewes and a pure-bred ram, all of one breed—this is the best way. It is not always possible to secure good young ewes. The price one would have to pay for them is sometimes prohibitive.

Old Ewes

Sometimes the beginner can buy a few old ewes and by giving them good care prolong their lives and raise some valuable lambs. This is a less expensive way of starting than buying young ewes. This way also requires more care and is more uncertain. Sometimes well-bred ewes that are too old to live with the rest of the sheep, with common farm care will live long enough to produce two or three lambs if given a little extra attention.

Starting With Lambs

Sometimes the owner of a large flock will sell the small lambs cheaply at weaning time, and the beginner can secure a few small ewe lambs from a good breeder and by giving them good care develop a fine bunch of ewes. A few small sheep placed by themselves and given the run of a farm will develop wonderfully.

Mixed Ewes

Many men have started with a bunch of mixed ewes and by using good rams and saving the best ewe lambs had in a few years a uniform, valuable bunch of ewes.

Cross Breeding

When cross breeding is practiced to produce mutton lambs it is better to have the ewes all of one breed and not keep the cross-bred lambs for breeders. This means buying new ewes to keep up the ewe flock or else every two or three years mating part of the ewes with a ram of the same blood and keeping the ewe lambs from this mating for breeding purposes.

How Old To Breed

Ewe lambs of the mutton breeds will mate soon after they are weaned from their mothers, and produce lambs when one year old or even younger.

If this is permitted it stops their development and stunts them in size.   Ewes should not be bred until they are one year old and some practical sheep men do not breed them until two years of age.

SHED BETTER THAN BARN BASEMENT
By Peter Tubbs, Seymour, Wisconsin

Have been keeping sheep for forty-five years.   I would advise keeping twenty-five sheep on 160 acres.   A cheap shed is better than a barn basement for shelter.   Keep them separate from cattle, horses and hogs.   They do not need much attention but the better you treat them the larger will be the returns.

WILL STAND ABUSE, BUT GOOD CARE PAYS
By Grover Krantz, Canal Dover, Ohio

A man that has had no experience had better start in the sheep business on a small scale, as no other stock requires the detailed care that sheep do to have them do their best.   But on the other hand, they will stand as much abuse and poor care and make more money than other stock.

**A Good Ewe Will Give More Milk for Feed Consumed Than the Best Dairy Cow.
Ewes' Milk Fed to Fat Lambs Will Produce More Income Per Gallon than Cows' Milk Sold in the Can or Made Into Butter.
The Lambs Do Their Own Milking.
The Ewe Has a Fleece and the Cow Has Not.**

PROFITABLE SHEEP MANAGEMENT

Plan For Keeping Bunch of Ewes One Year, Commencing November 1st

Up to November 1st the ewes can pasture in the different fields and lots.

If they are in good flesh and the pasture plentiful they can be kept in the Corn Belt up to this time without grain feed.

Light Grain Feed—Ear Corn

It is well, however, to begin about the middle of October, sometimes sooner, to give them a light grain feed once a day. This can be one-half ear of corn to each ewe, fed on the grass or in a wide, flat-bottomed trough, allowing the ewes to shell the corn themselves. Oats can be fed if they are grown on the farm.

The amount of grain to feed in the fall depends on the pasture and condition of the ewes.

Flat-Bottomed Troughs

All grain troughs for sheep should have flat bottoms. V-shaped troughs allow the sheep to get too much feed in their mouths at once and they will waste much of it when changing places at the trough.

Time to Breed Ewes

If early lambs are wanted the ewes will be bred before November 1st.

Ewes bred November 1st will begin lambing about April 1st.

It is the practice of many farmers to breed their ewes in November so that the lambs are born in April, when the weather is getting warm and the pastures are beginning to furnish grass for the ewes.

Advantages of April Lambs

It does not take as much feed or shelter for April-born lambs as for earlier ones yet there are good sheep men who claim that where there is warm shelter, plenty of clover or alfalfa hay and good silage to feed the ewes, the lambs are more profitable if born in February or March.

Advantages of Early Lambs

In Winter and early Spring the outdoor farm work is not so urgent. More time can be spent caring for the lambs. Early lambs learn to eat hay and grain before they are turned to pasture, thus getting a start ahead of later lambs. This makes them better able to withstand parasites and they can be fattened for earlier and better markets.

Breeding the Ewes—Care of the Ram

If there are only a few ewes and the ram is strong and vigorous, he can be turned with the ewes at breeding time and left with them for one month. If there are more than fifty ewes the ram should be kept away from them at night or a part of each day and fed and given water in a stall or lot where he cannot see the ewes.

Feed Box For Ram

The ram can be fed while running with the ewes by giving him some grain in a small box so that the ewes cannot steal it from him.

Take Ram Away From Ewes

The ram can be allowed to stay with the ewes until winter when he should be taken away from them. If he stays with the ewes through the winter he will eat too much, get "bossy," and bunt the ewes about, often causing abortions.

Shelter From Storms

When the weather begins to grow cold in October and November sheep should have shelter from rains. The important part of this shelter is the roof. It can be an open shed on one or more sides and there should be a fence or door to keep the sheep in during rains as they will not always go in out of storms, especially if the storm comes at night after they have selected their place to lie.

Sheep that are out in soaking fall or winter rains followed by cold weather get catarrh or "snuffles," suffer and lose flesh.

Ewes in Stalk Fields

Ewes can be turned into stalk fields after the corn is harvested. If the husking is carelessly done, leaving many ears in the field, there is danger of the ewes getting too much corn, causing founder. If the corn is fairly well gathered there is no danger. The ewes will not need grain fed to them while they can get ears in the field.

Turnips, Rape and Rye

Turnips, rape, soy beans or vetch sowed in the corn at last cultivation will often furnish much fall and early winter feed.

Early-sowed rye makes good fall, winter, and spring pasture.

Sheds and Hayracks

As winter approaches with hard freezing and snows, sheds and racks for feeding hay must be provided.

Silage and Alfalfa for Ewes

There is no better roughage for sheep than alfalfa hay. Next to it is clover, soy bean and pea hay. Early-cut oat hay is excellent roughage for ewes.

An ideal winter feed for breeding ewes is from three to five pounds per day of good corn silage to each ewe, and what alfalfa hay they will eat up clean. The silage should be made from well-matured, well-eared corn. If there is no alfalfa, clover or protein roughage they should have some oil meal, cotton seed meal, wheat bran, or a mixture of these to furnish protein. Corn fodder can be fed to the ewes in racks, shredded or cut, or whole stalks can be scattered on the frozen ground (better out on the pasture), where they can pick the blades off.

Clean Feeding Ground

Never feed fodder or hay day after day in the same place on the ground. Sheep do not like to eat from a place where they must muss over the feed.

Bad Practice of Feeding Fodder Day after Day in the Same Place. Sheep Will Not Eat Feed They Hava Trampad Over Unless Driven To It By Hunger.

Shock Corn

When silage is not available, breeding ewes can be wintered up to near lambing time on shock corn and alfalfa. The ewes can be fed the shock corn on frozen ground or pasture when the weather is not stormy. Feed the shock corn sparingly until the ewes learn to husk it themselves—then they can be fed enough shock corn to make one large ear to each ewe per day. The ears are easily counted as the shock corn is scattered, always allowing a few ears extra if it is a large bunch of ewes.

Feed the alfalfa in racks at evening.

There are numerous combinations of feeds for ewes. The feeds to be fed depend upon the feeds that you can grow on your farm and the kinds that you can buy the cheapest, providing they are good and suitable feeds.

Ewes Must Have Protein—Grow It

Breeding ewes should not be allowed to become thin in flesh. They should always have protein feed in winter such as alfalfa, clover or bean hay, oil meal, wheat bran, brewers grains, or gluten feed, to maintain their muscle and blood supply and to develop the unborn lambs. Cotton seed meal is a good protein feed and can be fed safely with silage.

It is always best to grow the clover or alfalfa to provide protein and not buy too much expensive feed.

Lambing Time

Watch the ewes closely at lambing time to see that the newborn lambs do not get lost from their mothers and that they get the first mess of milk promptly—after that, they will stand considerable cold and will look after themselves in a surprising manner if they have good mothers. ***Don't neglect them.***

(Photo from Oklahoma Experiment Station)
**Creep for the Lambs by Means of Which the Lambs may be Fed Separately
From the Older Sheep**

Hurdles and Pens

It is a good plan to have hurdles six to eight feet long made of light boards or lath, which may be set across corners of the sheep house to make separate pens for the ewes while their lambs are young. The hurdles can be tied in place with string or wire. Separate pens are especially important with ewes that have twins, to keep the ewe and lambs together, so that the ewe will not disown one of the lambs.

Keep Hogs and Stock Away

Keep hogs and other stock away from the ewes and lambs. Hogs will eat the lambs while they are young and horses and cattle will tramp and injure them if confined in the same yards or sheds.

Shear Early and Dip After Shearing

Early shearing is always advisable if there is shelter for the ewes on cold nights and from rain.

If there are ticks on the ewes dip both ewes and lambs right after shearing. *Be sure to dip the lambs* as the ticks will move to the lambs after the ewes are shorn.

Ewe Like Dairy Cow

Ewes should be fed like dairy cows if they are to produce a lot of milk.

Don't feed a ewe much grain for a day or two after the lamb is born—then her feed can be increased until the lamb learns to eat hay, grain, and grass, when her feed can be reduced and more given to the lamb.

Creep For Lambs

Build a creep for the lambs (a creep is a panel of fence with

This Feeding Trough Keeps Lambs from Jumping Into Their Feed

slats or palings far enough apart to let the lambs through but not the ewes), and place a feeding trough inside the creep in which the lambs can be fed.

Lamb Feeding Trough

Always make a lamb trough so the lambs cannot jump into it and soil their feed (see illustration on preceding page). Feed for the lambs at first can be wheat bran, cracked corn, a little oil meal, or any clean ground feed until they learn to eat. Then they can be fed cracked corn and oats, oil meal and silage, or any good feed or combination of feeds.

Keep Lambs Out of Hay Racks

Fix the hay racks for both ewes and lambs so the lambs cannot get upon the hay with their feet (lambs delight in climbing into racks). Give the lambs choice bits of alfalfa or clover hay. They will soon learn to eat silage.

When to Sell Lambs

A fat lamb at weaning time (lambs should be weaned when four months old), will usually bring as much as it will two months later and often as much as it will bring after being fattened in the winter. It is the practice on many farms to sell the lambs right off the ewes; others keep the lambs to fatten and sell during the winter; others do not sell the lambs until they are one year old, thus getting one fleece from them. Shearing is advisable with the fine wools as they produce heavier fleeces and do not mature quite so rapidly as the mutton breeds.

Keep Best Ewe Lambs

Keep a few of the best ewe lambs each year and discard a few of the older and less useful ewes.

Summer Feed

When lambs are to be sold at weaning time it is usually profitable to feed them grain until they are sold. This can be done by building a pen or creep in the pasture in which to feed them.

If the pasture is good and the lambs are fed grain liberally, feeding the ewes grain can be discontinued when the lambs are two months old.

Lambs grown for feeders, to be fattened during the winter need not be fed as much grain during the summer, as lambs that are to be sold at weaning time.

A Few Sheep Without Grain

Where there are but few sheep on the farm and there is plenty of range, fat lambs can sometimes be grown without feeding either ewes or lambs grain after the grass is plentiful.

Fresh Pasture Prevents Worms

If the ewes and lambs can have a fresh pasture every two weeks until the lambs are weaned, there will be little danger from stomach worms. Where there are only a few ewes on a farm this plan can be worked out.

If it is not possible to have change of pastures a close watch must be kept for indications of stomach worms. (See page 48.) By all means furnish the lambs fresh pasture after they are weaned.

Rape Patch on Every Farm

A small patch of rape is valuable on every farm for lamb and hog pasture.

(Photo Courtesy Pa. State College.)

This combination hay and grain rack, designed by Mr. Chas. W. Carothers, Taylorstown, Pa., keeps the sheep from wasting hay and protects the fleece from chaff and seeds.

The sheep must pull the hay through the slats in order to get it and the trough catches the leaves that shatter off, thus saving the best part of the hay.

With poorly made feed racks sheep sometimes waste as much feed as they eat.

Rape and sweet clover can be sowed with oats in spring and after oat harvest the rape and clover will furnish excellent pasture.

Lambs in Cornfield

If the lambs are not sold at weaning time there is no better place for them than in the cornfield to eat weeds and lower blades of corn.

Provide Water

Always arrange so that both ewes and lambs can get pure water. Sheep will get along on pasture without water but they will do better with it.

Salt and Tobacco

Sheep and lambs should have salt where they can get to it the year round. It will do no harm to have tobacco stems where the lambs can always get them; they will help prevent stomach worms.

Let the Ewes Clean Up the Farm

After the lambs are weaned the ewes can be used to clean up lots and fields on any part of the farm.

From August to November they can have the run of meadows, stubble fields, and wheat fields until the wheat comes up, eating volunteer grain and the grass and weeds along the fences. They can spend a few days in each of the barn lots. It will not harm mature ewes to occasionally confine them on weed patches until they clean up the weeds thoroughly.

Docking and Castrating

Big healthy lambs can be docked and castrated when three days old, but it is safer with average lambs to wait until one week old. The earlier this work can be done with safety the less will be the shock to the lambs. It is not necessary to dock and castrate the lambs if they are to be sold before they are three months old.

Always Disinfect

Always disinfect tools, hands, and wounds when castrating and docking lambs. There is nothing better for this than reliable sheep dip, mixed with warm water according to directions.

Cut Tails With Sheep Shears

Young lambs' tails can be quickly cut off with a pair of sharp sheep shears with stiff springs. Let the operator stand astride of the lamb, holding it between his legs with its head behind him. With the thumb and finger of left hand pull the skin at the root of the tail toward the lamb's body, and snip off the tail, leaving a stub about one inch long. When the skin is released by the left hand it moves back and helps to cover over the wound. *Always disinfect the hands and shears before operating.*

Have the disinfecting fluid in a large pan and lay the shears in the pan while not using them; also dip the hands in the pan.

Castrating

A pair of sharp shears with a stiff spring are better than a knife for use in castrating young lambs.

Keep a close watch on the lambs after docking and castrating that maggots do not get in the wounds.

Sore Eyes

Young lambs sometimes have sore eyes. Don't neglect them. Wash them with a mild solution of sheep dip.

Often sore eyes are caused by the eyelid's rolling in and irritating the ball. This trouble is frequent with Merinos, both lambs and older sheep.

This can be cured by nipping out with a pair of sharp shears a "button hole" about one inch long and one-quarter inch wide in the skin above or below, whichever lid it should be.

Cut close to the lid but not too deep.

As the "button hole" heals the skin is drawn back, pulling the lid open.

WARM WEATHER BRINGS MAGGOTS

Maggots hatch from the eggs of the blow fly. Blow flies are apt to lay their eggs in the wool on any part of the sheep's body if the wool is kept wet by continued rains.

Keep the sheep tagged or trimmed so no manure will get on the wool. Watch the sheep closely for maggots. Any good sheep dip will kill them.

SILAGE FOR SHEEP

Good silage made from mature corn is a safe and dependable sheep feed in the hands of a careful feeder.

Sheep on Farm of O. C. Shaffel, New Castle, Indiana

Sour or mouldy silage will kill sheep.

Lambs can be fed up to two pounds of silage daily and older sheep up to five pounds. Some good sheep men feed ewes suckling lambs, all the silage they will eat. Clover or alfalfa hay should be fed with silage. The amount of grain fed with silage depends on the amount of corn in the silage.

The writer has wintered breeding ewes and fattened lambs successfully on corn silage, made from mature corn with large ears on it, and alfalfa hay, with no other feeds.

Start the sheep on silage by giving them very little at a time.

Don't give the sheep a big mess of silage the first feed after they have been away from it a while.

Be careful not to feed mouldy or spoiled silage to sheep.

Don't feed the silage from around the silo doors, no matter how good it appears to be.

HOW TO CARE FOR BREEDING EWES
By C. L. Freed, Lancaster, Ohio

My grain feed consists of two parts oats, one part corn. Breeding ewes are also fed some bran and oil meal. I take good care of my sheep and lambs at lambing time. Protect the ewes and lambs from cold winds and draughts, and arrange their quarters so that no lamb can get out under or fast in anything. I usually have pens 3½ x 4½ feet long for the ewes with young lambs and twins. Shear all tags well away around udder before lambing. See that the young lamb drains both sides of the udder, and when there is an orphan carry it along with the use of bottle and cows milk until a foster mother is available for it. Put it with her in a pen and make her own it.

BE ON THE JOB AT LAMBING TIME
By John Foster, Williamsburg, Ohio

Lambing time is a very critical time and a man must be on the job and know what to do. There are many little things to learn connected with the sheep business. Sometimes a ewe will have twins and, feeling sore and sick, has a desire to walk off and leave them. Your attention is needed to gather the little fellows up and place them in a small pen with her so that they are separated from the flock. Nine times out of ten she will raise the lambs. I have saved many a pair of lambs that way. Another thing which happens very often is that a wax forms in the end of the nipple and the little fellow has not the strength to pull it out.

ONE EAR OF CORN AND PASTURE FOR EWES
By W. D. Spence, Fairbury, Illinois

Start with ten good mutton ewes and a pure-bred ram. About August 1st I begin feeding each ewe an ear of corn a day on pasture. This seems to make them mate earlier and my lambs arrive about February 1st.

After the ewes are bred I let them clean up stubble fields, fence corners, and weed patches, until after corn husking when they go to the stalk fields, where they stay until almost time for the lambs to come. Put them in a shed in stormy weather. Give them good care during lambing and feed both ewes and lambs a little grain.

HOW TO COMBAT SHEEP PARASITES
By H. J. Renk, Boise, Idaho

Saving the lamb crop is the key to success or failure. The ewes should lamb early and wean early. Put out on fresh succulent feed that has not been pastured by sheep before, to dodge the stomach worms and other internal parasites which take such a large toll of native lambs annually. Sow about two pounds Dwarf Essex rape seed and a few pounds of red clover and turnip seeds with grain crop every spring and you will have a fine fall pasture for lambs.

A thin, weak sheep or lamb is very susceptible to internal parasites or any disease. A great many diseases are warded off by nature when a sheep is in good flesh and has abundance of exercise. Never allow sheep to get thin or run down. In hot weather be on the lookout for blow flies, as maggots will hatch from their eggs in a few days and begin eating the host up alive. Use coal tar dips diluted in water.

KEEP UP STOCK WITH EWE LAMBS
By Clark James, Princeton, Illinois

I will buy a carload of young western ewes and use registered Shropshire bucks. Will dip ewes twice for ticks when received, and once every year both ewes and lambs will be dipped after shearing. By keeping 100 ewes I can ship a car of sheep every fall by keeping up the old stock with ewe lambs and culling old ewes with spoiled udders.

Unless I have a carload or part of one and divide the car I fill in with hogs. It's hard to dispose of a few. If only ten or fifteen sheep are wanted to keep down weeds, etc., home consumption and the local butcher can handle the surplus.

RAPE WITH OATS FOR SHEEP PASTURE
By J. S. Bumgarner, McNabb, Illinois

I sow rape with my oats and after harvest can carry from three to five ewes and lambs per acre until time to plow in the fall. I usually get from two and one-half to three months pasture in that way. I wean the lambs and put them in a cornfield in which rape was sowed at last cultivation. I also sow some rye early for fall and early spring pasture. If lambs are to be sold by June 1st they need not be castrated. If they are to run until fall by all means alter and dock them.

HOTHOUSE LAMB BUSINESS PAYS

By Wendell P. Miller, Fairlands Farm, Sunbury, Ohio

Hothouse lambs have furnished an important part of the winter meat supply of the larger cities for the past 25 years. When we first heard of the growing demand for these winter lambs it seemed to us that Central Ohio was too far from New York to safely send fresh meat without refrigeration, also it looked like an expert butcher would be required. Both these fears however were without foundation, for the express trains carry the meat safely and the butchering is very simple. Twenty years ago 10-week-old lambs sold for $8 to $10 each from Thanksgiving to Easter, while last year we sold several at $15 apiece and the market demand was not filled.

The Best Breeds

In starting the hothouse lamb business the first requisite is to get a flock of ewes that will breed in the spring. We prefer Merino-Dorset ewes and Tunis rams.

The lambs from this cross fatten early, the dark faces and short wool add much to the appearance, and the well-fattened carcass has just the proper pinkness of flesh to suit the New York trade. Another point in favor of the grade Tunis is the heavy development of the caul fat, which adds greatly to the finished appearance of the carcass.

May is the favorite month with us for breeding. If possible it would be best not to have the lambs born until the flock is ready to go into winter quarters, for in this way they would entirely escape stomach worms and can be gotten to eating grain much younger.

Best Feed for Breeding Ewes

If the pasture has not been abundant we begin to feed the

Ewes and Hothouse Lambs at Purdue Farm, La Fayette, Ind.

39

ewes about two weeks before lambing starts. Corn and some protein feed usually make up the ration. Remove mothers from the main flock. In case of Dorset or grade Dorset ewes with only one lamb it will be necessary to watch the udders for the first week and remove surplus milk. After that feed the ewes to give all the milk possible. Lambs whose mothers do not give sufficient milk will soon learn to utilize this excess if the ewe is held for them. When the lambs are two weeks old we put cracked corn and bran before them and their troughs are never empty except for cleaning three times a day. Oats, barley, or gluten is often added to give variety and stimulate the appetite, for we want the lamb to eat as much as possible at this early age. Bright clover, alfalfa, or soy bean hay is as necessary for the lambs as for the ewes, but the lambs should have theirs changed several times a day, as they will eat only the finer stems and leaves.

Silage Good for Hothouse Lambs

Clean water and salt must be accessible to the lambs as well as to the ewes at all times. Corn silage has been the essential part of the ewes' ration since we commenced raising hothouse lambs. It is fed twice a day and supplemented with oats, cottonseed meal, gluten, distiller's grains, or a mixed feed. A light feed of hay is given along with each feed of silage. The lambs are fed in a room shut off from the ewes by a creep, the slats of which should be just eight inches apart. When the lambs pass through this fence with difficulty they are big enough to butcher, the live weight being 45 to 55 pounds.

The uninformed person is often led to think from the term "hothouse" that the lambs are raised in an artificially heated building, but this is not the case as the only protection needed is that against cold, wind, rain and snow. On pleasant days the barn should be opened as much as possible and the air should be kept very pure at all times. Sunlight is especially desirable in the "lamb parlor" as we call the creep. The barn is kept clean with a light bedding of straw each day. Let nothing disturb the lambs or their mothers.

How to Dress Lambs for Market

The method of dressing the carcass has always been similar to hog dressing. At present only the stomach and attached organs are removed, the liver, lungs, heart and kidneys remaining untouched in natural position. The head is not removed until late in the spring. The preparation for market requires some skill in the finishing of the carcass. It is very important that they be thoroughly bled out by sticking the vein and artery in the neck close behind the head with a sharp pointed knife. Make sure that the artery is severed by noticing if the blood is bright red in color. For ease in dressing we suspend the lamb by the hind feet with a loop of binder twine around each leg, from a short single tree or stick bolted to the rafters. The belly is shorn closely from the brisket to the tail and up the inside of hind legs. Wipe the skin on the flanks clean with a dry piece of burlap. Open the lamb from tail to brisket and as the stomach falls out remove the caul fat and place it where it will remain warm. In cold weather place it in warm water. Back-set sticks are placed diagonally across the back to spread the carcass open and hold it firm.

Dressed Lambs Should Be Attractive

Carefully spread the caul fat over all the exposed flesh and pin it in place with a good grade of hardwood toothpicks. Make small slits in the caul over the kidneys and pull them through. It is here that care must be taken to make the carcass look attractive.

Hang the finished carcass in a clean room to cool for 12 to 24 hours, usually overnight. In the morning a small square of muslin is placed over the exposed flesh and sewed in place with strings across the back. Then burlap is wrapped about the entire carcass if the lambs are to shipped singly or the crate is lined with burlap if sent in a crate.

Part of the time we ship in small crates holding four lambs packed tight, and part of the time loose with only the burlap for protection. This will depend upon the commission man you send to and the number of railroad changes necessary. For the commission men the lambs should arrive in New York not later than Friday morning, preferably earlier in the week.

The first quotations on the New York market are made just before Thanksgiving. Commission men give the regular shipper careful attention. In some cases it is possible to arrange for a private trade, which will make the business more profitable. The demand for lambs continues active until after Easter.

HAVE CREEP FOR LAMBS
By D. D. Clifton, LaRue, Ohio

About one month before lambing time tag the ewes by trimming off the dirt and the locks and wool from around the udder, so the lambs will have no trouble in nursing. When the lambs begin to come stay with the flock as much of the time as can be spared from other duties, for some of the ewes will need help and some of the lambs will be weak and will need help to get their first milk. After nursing once they will usually take care of themselves.

When the lambs are about two weeks old they should be trimmed and docked. We have found the heated docking pincer the most satisfactory instrument to use. After the lambs are two weeks old make a creep where the lambs can feed at their leisure and the ewes cannot get in, and keep equal parts bran and ground oats before them all the time. Let them have all they can eat.

SHEEP DIP CURES STOMACH WORMS
By C. L. Robb, Cadiz, Ohio

We have used sheep dip for stomach worms and prefer it to the gasoline treatment as it is less severe on the sheep and we think it a very efficient remedy. We give about one teaspoonful of dip mixed with two-thirds pint of water. Older sheep may be given half teaspoonful more. Like all remedies it is best given after fasting from twelve to sixteen hours, and should be repeated in a few days.

We have also used the dip mixed with salt and kept before the sheep, using from two to four tablespoons to a gallon of salt, with good results. This also tends to keep flies from the sheep's nose. We never lost a sheep from using the dip treatment, but care should be used when giving any treatment not to strangle the sheep.

INCOME ALMOST FABULOUS
By L. B. Eidmann, Mascoutah, Illinois

I sold lambs about the 25th of March that brought $8 per head, weighing a little over forty pounds. Several of my ewes raised two this year which sold for that figure. The gross income per ewe would be almost fabulous. The fleece netted me over $4. I cannot say if all corn belt farms should have sheep. So many farmers are such careless stockmen that no stock will do well in their hands.

TURN HOGS AND LAMBS INTO CORN FIELD

Corn Belt Farmers Should Turn Them into the Field in August

The cost of harvesting a corn crop is as great as the cost of growing it.

It is a good plan to sow some rape and soy beans in the corn at last cultivation. It would be fine to have a strip of both rape and soy beans along one side of the corn field.

When the corn is full grown, turn lambs into the cornfield to eat the lower blades and weeds. The lambs will get an occasional ear of corn. By the time the corn is ripe the lambs will be on full feed of corn that they have pulled down. Now turn hogs in with them and let them all fatten together.

The lambs will soon get used to the hogs and will eat much of the corn that the hogs break down. By this method the gain on the lambs is almost all extra profit.

The weeds in the corn fields and the lower blades of corn that go to waste in the corn belt would make thousands of dollars' worth of high grade mutton, if good lambs were turned into the corn fields in August.

Lambs Eating Weeds and Lower Blades of Corn.

LAMBING OFF CORN WILL SAVE LABOR

By J. Orton Finley, Knox County, Illinois

My lambs and ewes did a splendid job of harvesting corn last year. They not only ate all the leaves off the stalk and all the corn off the cob, but they polished the stalks as well. We pastured off nearly 75 acres of corn in our fields with lambs, ewes, and a few shotes, and I am frank to say that my feeders have never done better. At night the ewes and lambs came into the yards for a feed of silage, alfalfa, and some cottonseed meal. They never went into the field when muddy.

Saved $200 in Labor

When it came time to husk, labor was scarce and prices high. I bought a half mile of wire fence and gave them ten acres at a time, the lambs going over the field first, followed by the ewes and shotes as the lambs passed on to a new ten acres. This saved me $200 in husking and extra labor in feeding. Kept moving the fence back last season until 75 acres were harvested, and I expect to do the same again this season.

[Mr. Finley's lambs usually top the market. He was the first man in the world to receive $11.15 a hundred for a car of lambs. A lamb never leaves the farm until his ribs have been touched to determine the degree of fatness. If he is not fat that lamb stays at home. Uniformity of breeding, finish, and quality is one reason why Mr. Finley's lambs are always market toppers.]

Sheep Eating Corn Leaves that Would Otherwise go to Waste and at the Same Time Cleaning up the Weed Seeds

LAMB FEEDING PROFITABLE WAY TO MARKET FARM FEEDS

There Are 5,000 Corn Belt Farmers Who Should Be Growing Alfalfa and Feeding Their Corn to Lambs Instead of Hauling It to Elevators

By Roscoe M. Wood, Douglas, Wyoming, and Saline, Michigan

Feeding lambs on the farm is a practical and profitable method of marketing much coarse feed which would otherwise bring little return to the farmer. This business is a development of the last twenty years and in that time its character has changed much.

Two Kinds of Lamb Feeding

Lamb feeding is practically of two kinds: The pasture and cornfield proposition of late summer and fall and the winter feeding in sheds on hay and corn. Whether a farmer pursues one or both systems there are several factors which apply with equal force.

We Must Like the Business

First—A man must have a liking for handling sheep. We read of occasional amateurs making phenomenal profits with their first bunch of feeding lambs, but these are like the prospectors who discover a fabulously rich mine—we do not hear of the men who lose. To fatten lambs requires observance of their peculiarities and humoring of their appetites. The same person should do the feeding—they should have their feed at regular intervals—both feed and water must be clean. Other animals should not have access to their pasture or feed lot, nor should any disturbance be permitted to excite them.

Western Range Lambs Best

In feeding the first thing is the lambs. Western range lambs are generally better than natives. They are free from disease. One lot runs of an even grade and they handle better in sizable flocks. Lambs weighing 52 to 60 pounds showing good bone and feeding capacity are most desired. Lambs carrying some Merino blood are hardier, make better gains, and are freer from death loss. These lambs can be bought in car lots through a reliable commission house at one of the leading markets or of a local dealer who may have bought on the market or direct from the range grower.

Be Careful in Feeding

However secured, they should be handled carefully when first brought to the farm. A timothy or blue-grass pasture is better than clover at first until the lambs are well filled and recovered from the shrinkage incident to shipping; or if put in the barn their feed should be limited until they have satisfied their hunger. Free access to salt should not be permitted. Given twice a week at regular intervals and scattered in the grain troughs is better.

Oats for Lambs

In accustoming lambs to grain we have found that putting oats in the troughs and sprinkling a small amount of salt over them will teach the lambs to eat grain most quickly, without a few getting too much. For this purpose oats are best as there is practically no danger of overeating. In a week's time corn can be added and the oats reduced and by the end of thirty days corn alone can be used. Many successful feeders, however, prefer during the entire feeding period, a mixture of oats, corn, and a little wheat bran, or dried beet pulp; linseed cake is also a profitable feed, but corn is the main grain. In beginning grain feeding it requires a quarter-pound per head per day, gradually increased to one pound per day and the last two or three weeks of feeding, all the grain the lambs will clean up at a feed even to $1\frac{1}{2}$ to $1\frac{3}{4}$ pounds per head. It is very import-

Lamb Feeding on an Illinois Farm. Note the Silos

ant that the increase should be gradual. Much loss has been occasioned by too sudden increase in the feeding of grain.

Alfalfa Excellent Roughage

For roughage clover hay has no superior. Alfalfa, where possible to secure, is excellent. In case of shortage of these feeds corn fodder, bean pods, oat straw, any or all will help out a shortage of clover hay, but some of the latter is essential. A large share of the profit in lamb feeding is secured by transforming these coarse feeds into marketable form and putting much of them back upon the land in the form of manure, thereby increasing the fertility of the soil.

Cornfield Feeding

In feeding lambs in the cornfield give them free access to it at all times after they are used to the pasture and surroundings. A good pasture in connection with the cornfield gives best results. Pure water, accessible at all times, is essential. Rape sown in the corn at time of last cultivation furnishes most acceptable roughage at a minimum cost.

Clean Racks and Troughs

Cleanliness, regularity and proper feeding are three important requisites to profitable lamb feeding. Hay racks must not be burdened with hay or roughage which has been picked and mussed over; grain troughs must always be clean and dry, no manure nor filth being permitted; water troughs or tubs must be likewise. A lamb's appetite is a perfect clock—he is hungry for his feed at the same time each day. Do not disappoint him by failing to have his grain before him when he wants it. Proper feeding consists in giving just the right amount each time and especially not too much. The lambs should clean up the grain troughs quickly and at the same time plenty should be given.

The lamb's appetite must be watched and humored. For instance, he will assimilate more feed in cold, dry weather than when it is warm and wet. The most successful feeder is he who watches his lambs and humors them most, for then they gain best.

STOMACH WORMS CAN BE PREVENTED

Change Pasture Every Two Weeks—Wean Lambs Early— Put Lambs in Corn and Stubble Fields—Feed Tobacco Stems the Year Round

Next to dogs, stomach worms are the most dangerous enemy of sheep and lambs. They do not always seriously affect older sheep, but hundreds of lambs suffer and die from this pest.

How Lambs Become Infested

The worms live in the fourth stomach. Mature sheep can look healthy, yet carry over winter enough worms to kill all the lambs the following summer. The eggs are scattered with the droppings on pastures where they hatch and the young lambs gather them up with the grass.

There is no danger in cold weather. As soon as the weather is warm the eggs hatch and if the lambs are kept on the infested fields, trouble and loss is sure to follow.

Symptoms of Stomach Worms

Lambs get droopy and dull—sometimes refuse to eat and lose flesh rapidly. Often there is swelling under the jaws. The bowels are loose and offensive and the skin gets chalky white instead of pink as it should be on a healthy sheep. Death often results unless effective and early treatment is given.

(Courtesy Lousiana Experiment Station.)
Lamb Infested by Stomach Worms

"Paper Skin" is the old name for the white, "papery" appearance of the skin of sheep suffering from stomach worms.

Prevention is the best cure. If the farm is fenced so the ewes and lambs can be changed to an entirely fresh pasture every two weeks until the lambs are sold, or weaned and turned into stubble or cornfields, there will be little danger from worms.

Tobacco Prevents Stomach Worms

Tobacco dust or stems chopped fine and placed where sheep and lambs can always have access to them will help prevent stomach worms. Instances have been reported where lambs that have had tobacco stems to eat remained healthy all summer on old pasture. There are worm powders and medicated salts on the market for which claims are made. Keeping tobacco stems before sheep the year round is a cheap and apparently effective preventive, but by all means arrange changes of pasture.

Drenching the Only Cure

It has never been shown that sheep will eat enough of anything to "cure" stomach worms. Anything that can be fed to them can be considered only as a preventive.

Drenching is the only cure when once the worms have possession and the sheep are losing flesh.

Here are three good treatments for stomach worms:

Blue Vitriol

Dissolve one ounce of blue vitriol in one gallon of water. Do this by putting the vitriol in a sack and hanging it in the water— it will not dissolve in the bottom of a vessel. Give this mixture in doses of from two ounces for lambs to three ounces for older

(Photo Courtesy University of Minnesota Experiment Station)
All These Lambs Were Same Age and Breeding—Little One on Extreme Left Was Infested With Stomach Worms—Others Were Protected by Change of Pasture

sheep after keeping them away from grass or feed overnight to allow the stomach to empty.   Repeat the dose in 10 days.   *Keep the mixture well stirred* so the last doses will not be too strong. Carelessness in mixing or giving this treatment is dangerous.

Coal Tar Sheep Dip

Mix the coal tar dips same strength as for dipping for ticks and give three ounces to lambs and four or five ounces to older sheep. Keep the sheep off feed overnight or 15 hours before giving the dip and repeat in 10 days.

Gasoline

Mix one tablespoonful of gasoline in five ounces of milk for lambs and one and one-half tablespoonfuls of gasoline to five ounces of milk for older sheep.   Keep the sheep off feed overnight before giving treatment.   Repeat in 10 days.

Mix each dose separately.   Be careful not to get too much gasoline.   Shake every dose well to mix the gasoline with the milk.

Give Treatment Three Times

It will make a cure more certain to give the worm treatment three days in succession, each time fasting the sheep overnight before giving the dose, then repeating in 10 days.   At the second treatment after the 10-day interval it is not necessary to give the three successive doses unless the sheep is badly infested.

Get Rid of Worms in Winter

Although it is not definitely known experience indicates that stomach worms do not live over winter in pastures. This gives a chance to get rid of worms in all the sheep during the winter and start clean in the spring.

A Self-Feeder Like This Filled With Salt and Tobacco Stems Kept in the Pasture Will Help to Prevent Stomach Worms

Drenching Ewes

Ewes can be dosed while in lamb if they are carefully handled.

After sheep or lambs have been treated put them on fresh pasture.

Never lay a·sheep on its rump or side to drench it. Always allow it to stand.

The safe way to hold a sheep for drenching is for the operator to back the sheep into a corner and stand astride of it.

Every farmer should have a metal drenching syringe. A large-sized syringe will do for cattle, horses, or sheep.

A long-necked bottle will do but there is danger of strangling the sheep by pouring too much into the throat at once.

Don't hold the sheep's nose high. Put the left hand under the jaw and hold the nose a very little higher than level, run the syringe or bottle back into the mouth but not far enough to pour the dose directly into the throat.

There will be no trouble in getting a sheep to swallow. Make three swallows out of an ordinary dose by pouring about one-third of it into the back of the mouth at a time and stopping long enough between times for the sheep to swallow.

There is great danger of strangulation if the head is held too high or too much is poured into the throat at once.

Sheep Should Have Shade. This Shed Could be Improved by the Addition of Three Sides to Darken It and Keep out Flies.

MUDDY YARDS MEAN SORE FEET

Have Dry Yards and Pastures—Foot Rot Dangerous

Sheep that are compelled to stay in muddy yards or in low, swampy pastures are almost sure to have sore feet. Foot troubles are given different names such as "foot scald," "hoof ail," "foul foot," and "foot rot," but all mean the same.

It is the belief of some farmers that there are two kinds of sore feet—the first called "foot scald," being the "scaldy" condition found between the toes when the sheep is first affected; and second, genuine "contagious foot rot" when the disease gets deep seated under the horn of the hoof.

There seems to be some ground for the belief that the two stages are different diseases since the first has been known to get well with no treatment except to place the sheep on perfectly dry footing. Such instances are so rare that the man with lame sheep cannot afford to take the chance of thus curing them.

Sore Feet Contagious

It serves the practical sheep man to treat foot diseases as contagious and needful of prompt attention.

The first symptoms are lameness and on examination the foot is found to be feverish and the skin in the cleft of the hoof red and swollen. Soon matter or pus forms in the cleft and in a short time, if left unchecked the disease gets under the skin and spreads under the entire horn or wall. Blow flies lay eggs in the diseased hoof during the summer months. The maggots spread from the foot to the wool, finally killing the sheep if left unchecked.

The thing for the farmer to do as soon as he discovers lame sheep is to get busy and stop the trouble before it reaches the advanced chronic stage. If the outbreak occurs in winter or spring the first thing is to provide dry quarters if possible. If it happens in summer put the sheep on the driest pasture obtainable until cured.

Doctoring Sore Feet

If there are but few sheep on the farm they can be caught and each foot treated separately. Treatment consists of cleaning all mud and filth from between the toes and applying something that will cure. In practical experience nothing has proven better than pulverized blue stone or blue vitriol (sulphate of copper), mixed with vinegar or water to a batter and applied with a paddle to the sore. If the lame sheep are neglected until the disease gets deep-seated all the horn must be pared away from diseased parts with a sharp knife so the remedy can reach the sore. The right way is not to allow the trouble to reach that stage.

Quick Way for Large Flocks

Where there are many sheep to be treated it can be best done by walking them through a trough containing a blue vitriol solution mixed at the rate of one-half pound of blue vitriol to a gallon of water. The trough can be arranged so the sheep must walk through it coming from the pasture to stable. This trough treatment is an excellent way to prevent sore feet.

Whitewash Preventive

Where there has been no lame sheep a trough containing a lime and water mixture about as thick as whitewash placed where the sheep will walk through it daily will prevent foot trouble. This is cheaper than curing after the sheep are lame.

Trim Feet Once a Year

The good sheep man will catch his sheep once or preferably twice a year and with "toe nippers" trim the long toes and cut away all excess hoof that will be likely to catch filth. Keeping the hoofs level adds comfort to the sheep.

Sheep Enjoy a High, Dry Place to Rest

CANNOT AFFORD TO FEED SHEEP TICKS

Dipping Kills Ticks and Cures Scab

Lambs with ticks on them do not grow.

Ticks cut down the ewes' milk.

Ticks suck blood from the sheep. The way to get rid of them is to dip the sheep thoroughly and repeat in about two weeks to kill the ticks that hatch after the first dipping.

Dipping can be done any time of year but best results are obtained in Spring just after shearing. Dip both ewes and lambs. Use any of the reliable sheep dips and follow directions.

Medium-sized sheep and lambs can be dipped in a large barrel. A home-made dipping box of wood or concrete is sometimes used, or galvanized dipping vats can be bought.

Dipping cures all skin diseases, keeps the sheep from "pulling their wool," cures sore eyes, and adds to the general health.

A thorough dipping, repeated in ten days will cure scab.

Keep the sheep in dip two minutes.

Be careful not to strangle the sheep while dipping by keeping the head under.

Sheep should be dipped every spring to prevent ticks, scab, and other skin diseases.

Grub in The Head

Inquiries about this trouble are frequent. Briefly, grubs in the head hatch from eggs laid in the sheep's nostrils by flies during the summer. The grubs work themselves up into the sheep's head and there is no positive cure, but they will come out themselves in their life cycle and change into flies to lay more eggs.

The farmer can prevent this trouble by providing dark places for his sheep to lie during the hot days of summer. Flies do not work in the dark, thus the sheep escape grubs and are comfortable out of the hot sun. A cheap shed can be darkened for them or they can be allowed to come to the barn and have a corner there.

Pine Tar on Noses

Pine tar smeared on the sheep's noses will keep the flies away. Where shelter cannot be provided tar should be used and it will do no harm to use both tar and shelter. The tar can be smeared on with a brush or paddle, or salt boxes or troughs can be arranged so the sheep will get tar on the nose while licking salt.

SHEEP BLOAT AND HOW TO TREAT IT

Bloating is a swelling that shows on the left side of the sheep's abdomen just in front of the hip bone.

It is caused by gas forming in the stomach from the fermentation of green leguminous feed, like clover and alfalfa.

Sheep are very apt to die from bloat if not relieved early.

Bloat usually happens when sheep are pastured on young and rapidly growing clover, alfalfa or rape.

There is more danger when grazing while the dew is on or after a rain.

Sweet Clover Will Not Bloat Sheep or Any Live Stock

There is little danger of bloating if the sheep are gradually accustomed to the clover, alfalfa or rape by turning them on it for only a short time each day for a few days before allowing them to have all they will eat.

It is best to turn the sheep into clover or rape the first time in the afternoon after they have had a chance to fill up on other grass so they will not eat so greedily.

If sheep that have become accustomed to clover pasture are taken away for a day or two and returned hungry, they are apt to overeat and bloat.

Care and good judgment will prevent practically all loss from bloating.

Remedies for Bloat

There are several simple remedies that if given in time, will relieve the sheep. One is to drench the sheep with a pint of milk after stirring in it a tablespoon of baking soda. If the milk cannot be had use warm water with the soda.

Another is to give a pint or more of milk fresh from the cow.

This remedy was first prescribed by Mr. Frank Kleinheinz, of Madison, Wisconsin, and has been successfully used since by many farmers.

Another: Give four tablespoons of linseed oil and a teaspoon of turpentine in a half pint of milk.

Relief sometimes can be secured by forcing the sheep's mouth open by holding a hammer handle or smooth stick crosswise like a bridle bit in the mouth and pressing gently on the swelled side.

It is always best to give one of the doses prescribed above as soon as the swelling is noticed and sheep is seen to be in distress; then the mouth can be held open. Sometimes a second dose in a few minutes will bring relief if the first one fails.

Pouring cold water on the swelling helps to cure, along with the other treatment.

If relief does not follow soon after these remedies are tried the last recourse is to "tap" the bloated side which is always the left.

This should be done with a trochar, but a knife will do.

Make a small incision at the highest most distended spot by running the knife blade downward, so as to puncture the stomach and allow the gas to escape.

It is best to insert a quill or reed, or small tube by the side of the knife blade to hold the incision open for a short time.

The Best Cure for Any Trouble is Prevention

Accustom the sheep to the pasture and don't allow them to gorge themselves on clover when hungry and there will be little trouble from bloat.

Merino Rams Owned by J. G. Helser, LaFayette, Ohio

HOW TO TELL SHEEP'S AGE BY TEETH

It is a great advantage for the beginner to be able to judge the age of sheep: it may save him from disappointment.

The age of sheep up to four years of age is easily told by the teeth.

Lamb Teeth (Fig. 1)

Young lambs have eight front teeth on the lower jaw. They are rather short and narrow and are called "milk teeth." There are no front teeth on upper jaw.

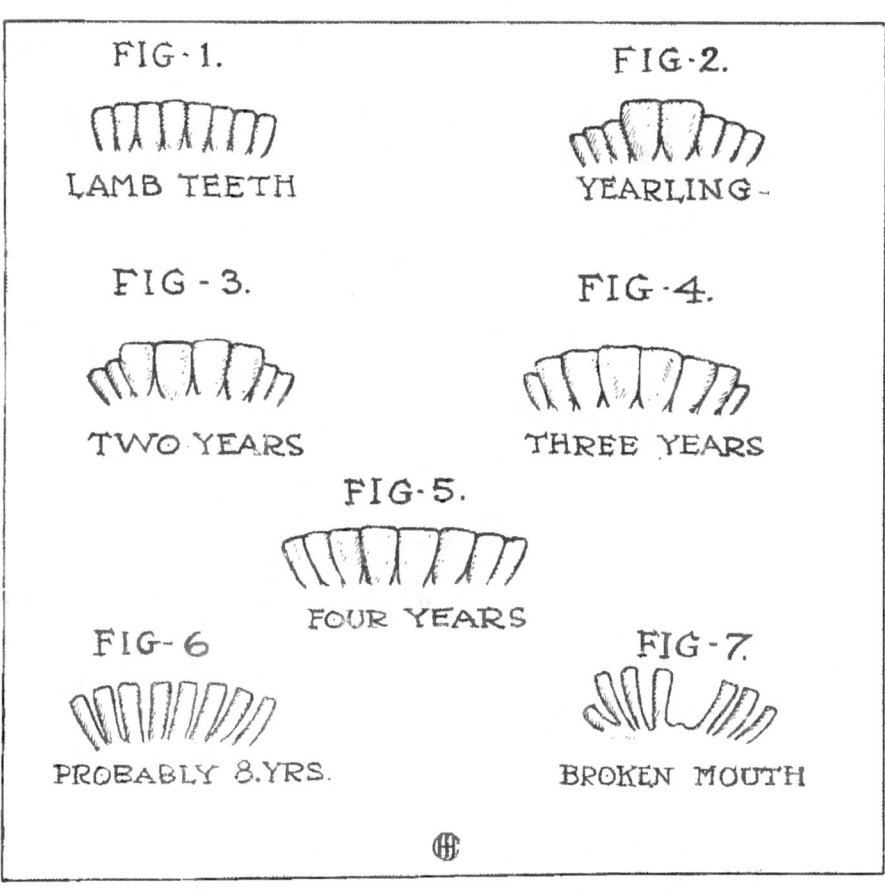

Drawing Showing Sheep's Teeth at Various Ages

One Year Old (Fig. 2)

When the lamb is about fourteen months old the middle pair of milk teeth drop out or shed and are quickly replaced by a pair of permanent teeth. These permanent teeth are much larger than the milk teeth.

Two Years Old (Fig. 3)

From twenty-one months to two years of age the next pair of milk teeth (the ones on each side of the permanent pair) shed and are replaced by another pair of permanent teeth.

Three Years Old (Fig. 4)

From thirty months to three years of age two more milk teeth drop out and are replaced by permanent ones.

Four Years Old (Fig. 5)

Within another year the last pair of milk teeth shed and give place to the last pair of permanent teeth. The sheep is now about four years old and has a "full mouth," meaning that it has all of its permanent teeth. For a year or two after the sheep gets a "full mouth" the front teeth change but little.

Getting Old (Fig. 6)

After the sheep gets a "full mouth" the teeth begin to slant outward, grow narrow, and get a "peg-like" appearance with space between them. Finally the teeth get loose and are lost one by one.

Broken Mouth (Fig. 7)

When some of the teeth are lost or broken the sheep is said to have a "broken mouth," and is considered past its usefulness.

Pulling Teeth

Breeders of pure-bred sheep sometimes pull the front teeth of valuable old ewes at this stage and by careful feeding prolong their lamb-bearing age. The loose, snaggy teeth are an annoyance to old sheep. They are better off with them out.

Difference in Teeth

Some mouths grow old in appearance much faster than others.
Sheep that are well fed and forced to early maturity often shed their teeth early and get "full mouths" before poorly-fed ones.
To determine the age after the sheep gets a full mouth is largely a matter of judgment.

SHEEP SHEARING NOT DIFFICULT

Shearing Machine Practical for 50 or More Sheep, if You Have Gasoline Engine to Run It

A letter in the National Stockman and Farmer by J. W. Hammond, Chief Sheep Investigator of Ohio Experiment Station, gives the status of the shearing machine for the average farmer:

Shear More Sheep

From the standpoint of speed the machine is decidedly superior to the hand shears. After a shearer has learned to operate the machine he can usually shear from 25 per cent to 50 per cent more than by hand. There also seems to be a fascination about shearing with the machine which makes it easier to get men to learn to shear by this method, than by hand.

Poor Work With Machine

In the hands of a slovenly workman the machine offers an opportunity for poor work but in the hands of a careful shearer it permits of better work than is possible with the hand shears. In machine shearing the sheep must be turned to suit the machine so that a different system of handling than that used in hand shearing must be followed, but a careful shearer will soon learn to handle his sheep so as to leave the fleece intact.

Don't Make Second Cuts

Another objection to the machine in the hands of some shearers is the matter of "second cuts" made. In shearing around the body there is a tendency particularly towards the end of the stroke to allow the shears to run out into the fleece instead of to hold them down close to the skin, making necessary the "second cuts" which shorten the staple, a very serious objection. A careful shearer will soon overcome this difficulty and not only leave his fleece in as good shape, but will do a smoother job than is possible with the hand shears.

Keep Knives Sharp

An essential to good work is to have the knives sharp and the tension properly adjusted. The tension screw should be turned a notch at a time until the proper adjustment is secured. A novice is likely to get his knives too tight, causing them to heat and do poor work.

Saving in Cost Doubtful

It is doubtful if machine shearing results in any great saving in cost. If one owns his machine, the cost of power, interest, depreciation, and repairs, will about offset the saving in time. However, when time is as valuable as it is on the farm in the spring it is worth something to get the shearing done quickly.

Don't Use Hand Power

The type of machine to buy will be determined by conditions. It is difficult to see any economy in the machine turned by hand, under most circumstances, because of its narrow cut, unsteady power, and the fact that it requires the time of one man to turn it.

Traveling Shearers

In some sections of Ohio there are men who own portable "two-man" machines and go around the country doing custom shearing at a fixed price per head. This seems to be a good solution of the shearing problem in the case of flocks too small to justify the purchase of a power machine and where enough flocks are kept to justify some one in the neighborhood in purchasing a machine. I doubt the economy of a farmer purchasing a machine of this type with a built-in engine merely for his own use.

Use Gasoline Engine

What looks like a better plan for a farmer having fifty or more

Hand Power — The Hard Way

sheep is to buy one or more shearing arms, depending on the size of his flock, operated on a line shaft driven by a gasoline engine or whatever source of power is used to drive the other farm machinery. Since hand shearers are becoming so scarce, it would seem that a farmer having a flock of fifty to seventy-five sheep and who already has the power to drive it, is justified in purchasing a machine of this type.

Good for Merinos

Because of improper construction and poor methods of grinding knives some of the earlier machines were unsuccessful when used on Merinos, but they have been perfected so that they are now as successful with Merinos as with any other class of sheep. For the last seven years we have used the machine on our experimental flocks of "B" and "C" type Merinos, shearing around 600 sheep annually. A few slight difficulties were encountered at first but since we have learned to operate the machine we would not go back to the hand shears. We have not used the machine extensively on very wrinkly or "A" type sheep but I can see no reason why such sheep cannot be shorn with the machine.

(Photo from Oklahoma Experiment Station.)
Machine Power — The Right Way

SHEARING SHEEP BY HAND

Hand Shears Still in Use in Many Sections Where Small Bunches of Sheep are Kept

Hundreds of small bunches of sheep will be started as farmers begin to realize their value. Men who start with only a few sheep sometimes have no farm power and there are not enough sheep in the neighborhood to warrant the purchase of a shearing machine. These men will commence by using hand shears. Men and boys who learn to shear with the hand shears will make good machine operators when the time comes.

These pictures and instructions, furnished by George B. Ford, Cambridge, Ohio, manufacturer of Ford Rigged Shears, are valuable for beginners and experienced shearers whether they use machine or hand shears. Mr. Ford gained a national reputation for rapid and high-class shearing in the Camps of Wyoming and Montana before the days of shearing machines. He could shear from 150 to 200 sheep in nine hours.

A great deal depends upon the different positions in which the sheep is held as to the shearer's ability in reference to speed and quality of work. The method shown here will, if carefully studied and practiced, prove a very great help to our eastern shearers in overcoming some of many difficulties encountered.

Do Not Grab Sheep by the Wool

Either catch them by the hind leg just above the hock, or by the flank. Never try to lead or move a sheep by pulling at its head.

Always put one hand under its neck and with the other push at the tail head.

Holding a sheep by the wool is apt to loosen the fibre from the skin, besides being painful.

Figure 1—Throwing the Sheep. Represents the shearer in the act of throwing the sheep. All that is necessary is to lift a few pounds rather quickly with the right hand, and the sheep falls right by the shearer's feet.

Figure 1

62

Figure 2—Opening up the Fleece. On straightening him up, assume the position shown in Figure 2. In this position the sheep is practically held by the shearer's legs. After shearing up the throat as shown, the sheep's nose is turned up toward the shearer's shoulder and should be held there until the side of the head and neck are sheared. Then the sheep's head can be controlled by the shearer's left arm while the hand is free to aid in the shearing.

Figure 2

Figure 3—One Side of Face and Neck Shorn. Now shearing down first side of body and around to backbone. After the foreleg and the shoulder are sheared as shown in Fig. 3, the sheared leg is placed behind the shearer's leg. This practically stretches the sheep's hide, gets the foot out of the way, and gives the shearer a nice surface to shear over. As shown in Figure 3, the shearing should extend around the sheep to within one or two inches of the backbone, but should not extend across it as it makes rough and uneven work.

Figure 3

One secret of skillful shearing is to always hold the sheep so that the skin over the part of body being shorn is stretched smooth—this prevents "nicking" or cutting the skin with shears.

Figure 4—Finishing the First Side. After shearing down the flank and having sheared the front part of the hind leg as shown in

Figure 4, place the left hand as shown in illustration. This will extend the hind leg so that you can shear from the hoof right up the leg and clear around to the back. In Figure 4 the sheep's head is practically free but kept under control by the left elbow. In going from Figure 4 to Figure 5 shear down over the sheep's ham to the tail. Shear the tail and nearly all that is between the points of the hips.

Figure 4

Figure 5—Shearing Up the Back. Making long sweeping cuts—shear a little past the backbone. In Figure 5 the sheep's breast and throat are held tightly against the shearer's leg.

This is done by catching the sheep by the unshorn wool near the right ear. The sheep is also held in this position by the right knee, resting lightly on the sheep's side, and the sheep can be drawn towards you with the knee as you shear over and up the back. Remember no extra weight is allowed to rest on the sheep in this position.

Another secret of good shearing is to have the sheep in position so that the wool falls naturally

Figure 5

away from the shears as fast as it is cut off.

Figure 6—Finishing the Head. After shearing up the back as shown in Figure 5, remove the knee and with the left hand straighten the sheep up and place right foot at sheep's back as shown in Figure 6. Here, as in all other illustrations, the wool is allowed to fall away from the shears of its own weight, while the left hand is employed in other ways. In Figure 6 the sheep's left front leg is behind the shearer's left leg. The sheep should not be held up straight but should lean lightly toward the shearer.

Figure 6

Figure 7—Rapid Shearing Down Side. After shearing down over the shoulder you assume the position shown in Figure 7. Here again you have an excellent opportunity to push your shears well into the wool and roll it off. Keep position shown in Figure 7 until you have sheared to the sheep's flank when the sheep should drop towards you just a little as shown in Figure 8.

Following this plan, the wool on the belly is left until the rest of the fleece is removed. The belly wool is then much easier and more rapidly shorn.

Figure 7

Figure 8—Nearing Completion. The best part of the fleece is now off—watch that the sheep does not kick into it and tear

Figure 8

it apart. In Figure 8 the sheep's breast and shoulders rest up against the shearer's legs. The sheep's head is held in position by the shearer's left elbow. In fact, the sheep's head is largely controlled by the shearer's left arm and elbow. As you shear on down in Figure 8 the sheep inclines more and more until his shoulders rest on the shearer's feet. Then after shearing past the position in Figure 8, the left hand is placed in the sheep's flank by pressing lightly, the hind leg is extended and the sheep's ham is brought up towards the shearer where he can easily finish the shearing of it.

Figure 9—Shearing the Belly. The fleece now being detached can be taken to the tying bench by the tyer or the sheep's feet can

Figure 9

be turned a little to the left away from the wool as shown in Figure 9. This is a very easy position to assume and after shearing down over the brisket as shown, the front legs are lifted by passing the left fore-arm under them. In no other position can a sheep's belly be shorn to as good advantage, especially if the sheep is poor or hollow from the lack of feed. Shearing on down as you come to the position in Figure 10.

Figure 10—Finishing. Trimming inside of hind legs. There is a great deal of time wasted by most eastern shearers by catching the sheep's foot and trying to hold it while shearing the leg. This should not be done as the sheep will invariably kick and struggle to get loose. Place the left hand in the sheep's flank, catching the hide, and by pressing a little against the sheep's leg will cause it to extend and you will have no trouble to shear from the hoof down to the sheep's body, then across and out the other leg by turning the sheep just a little to the right and press-

Figure 10

ing on the leg near the flank to keep it extended and to hold it steady while you finish trimming it.

Early Shearing Best

If shelter cannot be provided better leave the fleece on until weather is warm.

Sheep are very sensitive for a short time after the wool is removed. They suffer from cold rains and if compelled to stand in the hot sun when closely shorn, become sunburned.

Protect them for a few days. After that they will stand extremes of weather with very little discomfort.

Early shearing means tags and locks of wool are out of way so lambs can get to udders, lessens danger from maggots, saves the loss of wool from dung locks when sheep are turned on early pasture. Sheep shorn early make better gains.

MAKE THE FLEECES ATTRACTIVE

In tying or rolling fleeces observe the following: Have a table large enough to spread the fleece out full size like a blanket. Be careful in shearing and carrying the fleece to table to not tear it apart. A barn door laid on four barrels makes a good wool table.

How It Is Done

Turn the fleece on the table with white or flesh side down and spread out carefully—then fold in the ragged edges until if it is a large fleece, it is about twenty inches by three feet in size. Now begin at one end and roll, not too tight, or fold over into a neat bundle so that nothing will show except the white side. Pass the string directly around the middle of the fleece one way, then cross and round the other way—draw tightly and use no more string than necessary. This makes a square neat fleece that should be kept clean and carefully stored.

Tying boxes, illustrated below, are much used and the plan with them is much the same as outlined above.

Old Time Method of Marketing Wool Wrong

Selling small lots of wool on the farm has never been done in a satisfactory or business-like way.

The buying is done by small dealers who travel from farm to

(Photo from State College, Pennsylvania)
Fleece Rolled Up Ready for Tying

68

farm. They buy on commission for a larger dealer or manufacturer or else buy on their own hook, depending on selling the wool when they get enough of one kind together. Manufacturers will not look at small mixed lots.

All Looks Alike

Small lots of wool are not bought from the farmer on their merits, as often neither buyer nor seller are judges of wool. The country dealer buys as cheaply as he can because he knows the large dealer will throw out or "dock" in price part when he buys it. He also knows that some of the wool has manure wrapped up in it, some has burs, chaff, seeds, and straw mixed through it, and often there are two or three grades of wool in one lot.

When wool put up in bad condition reaches the manufacturer, it helps to give all the wool in that part of the state a bad name.

Wool manufacturers are human beings—they are apt to remember the burs, manure, and trash they got in last year's wool when they make offers on next year's clips.

DON'T SHEAR SHEEP WHEN WOOL IS DAMP
Tie Fleeces with Wool Twine—Remove Dirt and Filth
By Simon Summerfield & Company, St. Louis, Missouri

Always tie fleeces with regular wool twine, flesh side out. Never use any other twine and under no circumstances use sisal twine. Remove dung locks, chaff, seed, and all other foreign matter from your fleeces, and keep the different grades, especially burry and fine, separate from the other wool.

Keep Away from Stacks and Burs

If possible, keep sheep away from straw stacks, otherwise the straw will penetrate the wool, making it defective and discounting its value from 15 to 20 per cent compared with prices that could be realized for similar wool in merchantable condition. Sheep should also be kept away from burry patches as burs in the wool depreciate its value considerably. Sheep should be kept in green pastures, as the wool is always lighter in condition and will bring higher prices than wool shorn from sheep that are allowed to run at large.

Sheep should never be shorn right after a rain or heavy dew, for when wool is tied in fleeces while damp it is liable to damage the fibre and cause it to be classed as unmerchantable. Wool to be in merchantable condition must be thoroughly dry.

HOW TO PREPARE WOOL FOR MARKET

By S. Silberman, Boston, Massachusetts

The most injurious feature of wools raised in the middle west and eastern states is the large percentage of seedy and burry fleeces. It is important that the sheep be fed in such manner that the seed, straw, or hay does not get in the fleece; also sheep should be kept out of burry pastures, as should the fleece be either seedy or burry a great percentage of its manufacturing value is lost, resulting in at least a five-cent-per-pound reduction when selling on the market.

Twelve Months' Growth

It is preferable that the shearing be as close to twelve month intervals as possible, insuring a uniform length. Wools without sufficient staple—called clothing wools, are worth from two to three cents per pound less than combing wools.

Tie With Paper Twine

In shearing it is best that the fleece be retained in one piece and it is very important that it be tied, shorn side out, with suitable twine. A fibre twine and even the so-called wool twine, although the latter to a lesser extent, becomes entangled with the wool fibre itself, making it difficult to abstract it during the manufacturing process. These fragments do not have the same dye absorbent qualities as wool, so that they are plainly shown in the finished goods as streaks. This is one of the chief objections found with domestic wools as contrasted with foreigns. A hard finished or paper twine has been approved by manufacturers, and can be obtained from any of the large wool or twine dealers.

Mistaken Opinion

In conclusion we would like to correct an erroneous opinion of domestic wools. Well-kept Shropshire wool in this country is equal to that of any country and has superior qualities to South Americans, Australians, and many other localities. The grease value may not be as great but this is due to the skirting and greater shrinkage of many foreign wools. However, taking into account the number of pounds obtained per head the American grower can realize more for a good fleece than any other grower, if the wool is properly taken care of and the improved methods of breeding and care of the wool when shorn, be adopted.

HAVE SHEEP OF ONE BREED

The Quality and Length of Wool Will Be Similar—Neighborhoods Would Profit by Keeping One Breed

By Farnsworth, Stevenson & Company, Boston, Massachusetts

Our recommendation would be that care should be taken that the sheep on one farm should be as nearly as possible of one breed so that the quality and length of wool would be similar. This might be carried further so that farmers in one vicinity might raise wool of approximately the same grade, which would therefore be of interest to one buyer.

Different mills use wool of different qualities and if a farmer's clip is irregular it must be graded before it can be sold to advantage.

If the sheep are bred on the farm care should be taken that well-bred rams are used. The result shows very distinctly in the wool. It is bad judgment to economize here.

Make Wool Attractive

In preparing the wool for market the object to be obtained is to make it as attractive as possible to manufacturers, who are the ultimate consumers. Burs, chaff, etc., in wool injure its value and care should be taken in handling sheep to avoid this when it can be done.

Keep Tags Out

After shearing the heavy tags should be taken from the fleeces and kept separate and the wool should be tied with a small, hard twine, fibres of which will not unravel and become mixed with the wool fibres. Sisal, or fibre twine, is very objectionable, many manufacturers refusing to buy wool with which it is tied.

The importance of this is that twine is usually made of vegetable matter and takes dyes differently from wool, which is animal. Small pieces which may have been mixed with the wool go through the various processes of manufacture and are not detected until the goods are finally dyed, when they cause noticeable imperfections.

The United States Government has commandeered the wool clip of 1918, fixed the price and regulated the manner of handling the wool from the grower to the manufacturer. It is probable that this plan will be in force during the war and for a period afterward.

A GOOD WAY TO MARKET WOOL

Method Recommended by N. Gladys Dimock, Assistant Secretary of Otsego Sheep Breeders' Association

On March 2, 1915, a group of nineteen men met at Hartwick, New York, and organized the Otsego County Sheep Breeders' Association as an auxiliary of the Otsego County Farm Bureau.

Previous to the organization of this association the members had been selling their wool for just what the buyer who went from farm to farm would pay for it.

One of the first things taken up by the new association was the co-operative selling of their wool. Each man agreed to stand by a committee of three which was appointed to take care of the arrangements.

Sale Was Advertised

We advertised the sale which was to be held at this office and had samples of wool from our three largest flocks, two of which were pure-bred and one grade. We sold medium wool. Arrangements made before bidding began were that the price should be one-third less for cotted or fine wool. Sealed bids were not accepted, which was wise.

$1,000 Saved by Selling at Auction

The highest price offered just before the sale was 31 cents. The wool sold at auction for 36 cents, an approximate saving of over a thousand dollars. The wool was loaded at two towns, the farmer receiving his pay at the car.

The Second Year

The sale was conducted along the same lines the second year. The sale was widely advertised through a Boston commercial paper, and by sending out letters to some hundred buyers and manufacturers. This year we had a professional auctioneer who sold the wool for $39\frac{1}{2}$ cents, which was four cents over the market price.

Practically the whole wool clip of Otsego County was disposed of this year through this Sheep Breeders' Association. A letter was sent to the members asking them to consign their wool to the sale, the amounts ranged from 35 to 2,500 pounds. The smallest grower received as much per pound as the man with the large flock. Last year very little wool was discounted as fine or cotted. Two men from the Association were at the car to help weigh the wool at both shipping points.

Co-operative Selling Satisfactory

The members were very well satisfied. This co-operative sale is simply an object lesson, a stepping stone to sales of lambs, hay, and other products of the farm.

The Third Year

The following report of the third yearly sale of wool is copied from "The Otsego County Farm Bureau News":

The wool committee of our Sheep Breeders' Association, the members of which are H. H. Marlette, John A. Curry and Howard Cunningham, sold the season's wool clip at auction on July 11th to P. W. Talbott & Sons of Binghamton. C. W. Peaslee of Laurens successfully auctioned off the wool and secured the price of 69⅞ cents per pound for medium wool. One-third less is to be received for the small amount of rejects to be sold, which include fine, cotted, burry, seedy and black wool. More buyers were present than usual, there being men present from New York, Philadelphia, Binghamton, Utica and local points. Sales conducted for three consecutive years demonstrate not only the value of this sort of co-operation but also the fact that farmers can pull together on well-defined and feasible projects.

Will They Ever Learn?

We have often stated that farmers will co-operate as readily

Much Feed Is Wasted by this Method of Feeding. Chaff and Seeds Get Into the Wool, Lowering Its Value

as any other class of men but that due to certain conditions which exist it is harder for them to do so. Still there are some who are so "independent" that they continually lose money and the respect of their neighbors. This fact is well illustrated in this county by our sheep breeders. Those who are not in the association have sold probably 10,000 pounds of wool this year fcı from ten to twenty cents a pound less than that secured by the association, a loss to the county of not less than $1,500. Before the sale, buyers rode the county and talked so eloquently that they even persuaded a few members to sell in face of the fact that nowhere in the state was a price secured as high as that received by the association the two previous years. One member sacrificed not less than $80 by selling in this way. The association will make at least $3,000 for its members this year. Those growers who sold at private sale and secured more than fifty-five cents per pound can thank the association for every cent they secured. The agricultural agent of Ulster County stated that on July 9th the highest price being paid in that county was fifty-five cents. Buyers here realized that in order to get any wool at all they would have to pay more, so they offered as high as sixty-two cents shortly before the sale which realized $69\frac{7}{8}$ cents. No sheep breeder in this county should sell a single pound of wool next year except through the association.

Eventually Co-operation

The time is coming when men who have a bunch of sheep will carefully prepare their wool, bring it to a central point to be graded by someone who knows how, and the different grades sold to dealers or manufacturers.

This method is in practice in Canada under the direction of the Department of Agriculture.

On the western ranges sheep men are co-operating. Large shearing sheds have been built where thousands of sheep are shorn co-operatively, the fleeces "skirted" which means the stained and dirty wool from the legs and bellies is separated from the clean, even wool, and sold direct to manufacturers.

Do Your Best

If there are not enough sheep now in your county to have an organization, do your best in preparing your wool and sell it alone. *But talk to your neighbor. Help him to get some sheep. Talk to your county agent about sheep.*

WHICH BREED?

Start a Bunch of Sheep on Your Farm or Improve the Ones You Have

The breed you choose is not so important as to have good specimens of the breed. Any of the breeds do well in a small bunch. Merino, Dorset and Tunis ewes mate earlier in the fall than other breeds and for that reason are well suited to raising winter lambs for early markets. The large mutton breeds are heavy feeders and must have abundant feed. The Merino is hardiest of all and better adapted to hilly land and where large numbers are kept together.

The Medium Wool Breeds

Figure 1—The Shropshire is the most popular breed of mutton sheep in the United States. This breed is so well known that it needs no description. Its conformation is next to the Southdown in perfection. The Shropshire matures early and is almost completely covered with thick wool of medium length.

Fig. 1—Shropshire Ram, owned by Geo. McKerrow & Sons, Pewaukee, Wis.

Fig. 2—Southdown Ewe, owned by Chas. Leet & Son, Mantua, Ohio

Figure 2—Southdown sheep originated in Sussex, England, and are one of the oldest breeds. Southdowns are the nearest perfect mutton type; are short-legged, symmetrical, smooth, and compact and are deceiving in weight. They often weigh as much as some of the larger-looking, more upstanding breeds. Southdowns have brown faces and legs and medium weight fleece.

75

Figure 3—Hampshire sheep came originally from the Hampshire district of England. They have strong constitutions, size, and mature early. Rams should weigh when matured 300 pounds or more, and ewes 200 pounds or more. Flocks of breeding ewes average from 7 to 10 pounds of wool. Wool is medium length and of strong fibre. Hampshires have large heads with Roman faces. The face and legs are the blackest of any of the down breeds. Hamp-

Fig. 3—Hampshire Ram, owned by C. O. Judd, Kent, Ohio

shires have good mutton carcasses with strong, bony legs and large, open feet. The last year in which we have full statistics more Hampshires were imported into the United States than all other breeds combined. Hampshire rams are much used on the western ranges.

Figure 4—Oxford sheep originated in Oxford County, England, from a mixture of the Hampshire and Cotswold. Oxfords

are large. Rams weigh from 250 to 300 pounds and ewes 175 to 250 pounds; are broad and square in form and are rapid growers when well fed. The fleece is long and moderately open. Oxford sheep are very heavy boned. Head and face are somewhat like Shropshires, gray and brown in color.

Fig. 4—Oxford Ram, owned by Geo. McKerrow & Sons, Pewaukee, Wis.

The breed you choose is not so important as to have good sheep of that breed.

Good breeding without good feeding is a waste of time and energy.

Figure 5—Cheviot sheep originated in Scotland and for centuries have lived on the Cheviot hills in flocks of thousands, exposed to weather, the only feed being grass or hay. Rams weigh 225 pounds and ewes from 120 to 150 pounds. The Cheviot belongs to the middle wool class. Ewes shear from six to nine pounds. They are very prolific, a large percentage of the ewes raising twins. The fleece is pure white, faces and legs smooth, eyes bright, proud in carriage, and almost equal to Merinos as foragers.

Fig. 5—Cheviot Ewe, owned by G. W. Parnell, Wingate, Ind.

They do particularly well in large flocks, on the range and in rough, hilly country.

Figure 6—Dorset sheep originated in southern England. Both rams and ewes have horns and it is claimed they will defend themselves against dogs. They are very prolific. The ewes will produce lambs twice a year. Twins and triplets are very common. Dorsets have good mutton form and good quality of wool, but are not heavy shearers. They are excellent as producers of winter lambs.

Fig. 6—Dorset Ewe, owned by Park Ridge Farm, Park Ridge, Va.

The most successful sheepmen are the ones who chose a breed and then stuck to it. There is no good excuse for the sheep on any farm "running out" or deteriorating. Each crop of lambs should be better than the preceding one.

Figure 7—Tunis sheep are called fat-tailed sheep because they have broad, fleshy tails, sometimes six inches across. They

come from Tunis, in northern Africa. They are noted as mutton producers and the ewes will mate twice a year. This makes them valuable in producing winter and early lambs. Both sexes are hornless, have faces and legs bare of wool and covered with yellow-brown hair. Tunis sheep are fair wool producers, shearing from five pounds

Fig. 7—Tunis Ram, owned by Chas. Roundtree, Crawfordsville, Ind.

up of medium combing wool.

Tunis sheep are not as numerous in the United States as the breeds heretofore described, although their value as mutton producers is generally well known.

The Long Wool Breeds

Figure 8—The Cotswold came from the hills of England and is one of the largest breeds. They have a remarkable upstanding and stylish appearance with a good mutton carcass. The face and legs are white. Cotswolds are heavy shearers. The wool is very long and grows in ringlets with a long forelock hanging over the face.

Cotswold blood has been extensively used by western range men.

Fig. 8—Cotswold Ewe, owned by Alex Arnold, Gatesville, Wis.

Figure 9— **Lincoln** sheep originated in Lincolnshire, England. They are very large and the most compact and squarely built of the long wool breeds. In general appearance Lincolns resemble Cotswolds, except that they do not grow so much wool about the face and are of a much more rugged conformation. The face and legs are white. The face is long and bare with a tuft of wool or fore-top. The wool is long and coarse. Lincolns are great feeders, mature early, and make good mutton.

Fig. 9—Lincoln Ram, owned by Alex Arnold, Gatesville, Wis.

Figure 10—**Leicester** sheep originally came from Leicestershire, England. They are quite popular in Canada, but in the United States they are seldom seen. Leicester rams have been used in a limited way in range flocks. They are good mutton producers with rather fine, long wool, and bare, white legs and faces. The face and poll are entirely bare. Like the Cheviot, Leicester sheep are fine-boned and appear tall and leggy. The wool is not so dense as on the Lincoln and Cotswold.

Fig. 10—Leicester Ewe.

Figure 11—Corriedale sheep have just recently been introduced into the United States from New Zealand. They are a combination wool and mutton breed. Rams weigh up to 300 pounds and ewes around175 pounds. Ewes shear about 12 pounds per fleece and rams 20 pounds or more. The Corriedale originated from a mixture of Lincoln and Merino blood. They are very vigorous and hardy, are easily herded, develop rapidly and

Fig. 11—Corriedale Ram, owned by F. S. King, Cheyenne, Wyo.

lay on flesh almost as rapidly as the strictly mutton breeds. They have white faces and are covered completely over body with dense, long wool. Corriedales give promise of being a great range sheep and also practical for the average farmer.

There are only a very few Corriedales in the United States, they being located on the western ranges.

Figure 12—Romney Marsh or Kent sheep came from the marshes of Kent in the southern part of England. They have been bred for many years on the low-lying marshes. They are an excellent sheep for range conditions. The Romney is a long-wooled sheep, having a very heavy fleece. In appearance Romneys are very similar to Corriedales. Romneys have only been in the United States a very few years,

Fig. 12—Romney Ram.

but their numbers are steadily increasing. The Romney approaches an ideal combination for wool and mutton.

Figure 13—Karakul or Persian Fur sheep are natives of Turkestan and Persia. This breed is valued for the skins of the

young lambs, which are killed for their pelts when three or four days old. Fine quality pelts are worth as much as $12 each. Karakuls belong to the fat-tailed family of sheep. They mature to medium size and the mutton is of high quality. The rams are usually horned, the ewes are hornless. The ears are small and pendulous, the face narrow. Face and legs are

Fig. 13—Karakul Ram, owned by Middlewater Cattle Co., Middlewater, Tex.

covered with short, glossy, dark hair. The wool is long and hairlike and varies in color from light gray to black.

Mr. Alex Albright, Dundee, Texas, has developed the "Karaline Fur Sheep," which is a cross between Karakuls and Lincolns.

The Merino or Fine Wool Breed

The Merino is the hardiest of all breeds and produces the finest wool. Merinos came from Spain where they were kept in great herds and trailed many miles to pasture. They have been developed in the United States into three classes.

Figure 14—"Class A," very wrinkly and has a dense, fine, oily fleece. Rams of this type sometimes shear more than 30 pounds and ewes over 25 pounds. They are usually completely covered with wool on face and legs. "Class A" Merinos are not valuable for mutton, but produce the highest grade of clothing wool.

Fig. 14—Class "A" Merino Ram, owned by W. M. Staley, Marysville, Ohio

Figure 15—"**Class B**" **Merinos** are larger than "Class A's," have fewer wrinkles, and longer, bulkier fleeces, containing considerable oil. They produce a fair mutton carcass and at the same time grow very heavy fleeces.

Class "B" Merino rams are in demand by range men to use on light-fleeced range ewes. This cross increases the weight and quality of the lamb's fleece.

Fig. 15—Class "B" Merino Ram, owned by
R. A. Hayne, Adena, Ohio.

Figure 16—"**Class C**" **Merinos** are bred for a mutton carcass and have a fine long fleece of Delaine wool. They are very valuable for range and farm, as the Merino will thrive in large flocks better than any other and stand more hardships. "Class C" Merinos often have mutton carcasses equal to the best mutton breeds. Merino ewes of "B" and "C" types make good mothers for raising mutton

Fig. 16—Class "C" Merino Ram, owned by
C. L. Robb, Cadiz, Ohio

lambs when mated with mutton rams. The ewes will mate any time in the year, are long lived, and shear heavy fleeces.

Merino sheep have been developed and improved to their present state within the United States. They are the only sheep that importations from other countries are not made to secure sires and new blood.

They are also the only breed of sheep exported from the United States to other countries in any great numbers.

Figure 17—The Rambouillet is a Merino that has been developed to a large size, and bears a heavy fleece. The face and legs are completely covered with wool. They are noted for hardiness, longevity, and ability to make good use of feed in production of both wool and mutton. Many Rambouillet rams are used on the ranges, and are also a good farm sheep. In sheep shows they are also divided into classes, like other Merinos.

Fig. 17—Rambouillet Ewe, owned by F. S. King, Laramie, Wyo.

FEDERAL DOG TAX

There are probably 25,000,000 dogs in the United States, and 50,000,000 sheep. A Federal dog tax would be a great stroke of good legislation.

Let us have a Federal dog tax, make it high enough, and provide that the tax shall be collected in every case or a dog grave dug.

WRITE YOUR CONGRESSMAN.

A few million dog skins will help relieve the leather famine.

ANGORA GOATS
By H. S. Mobley

On my farm in the Ozarks in northwest Arkansas, I had a forty-acre tract lying on a hillside that was covered with rocks and boulders and shrub brush. It was as real a piece of waste land as you could find in any country.

In 1906 I fenced this forty acres and paid $15 for a small bunch of Angora goats. I have forgotten the number. Since that time I have sold out of that herd of goats $375 worth and have had goat mutton for the family whenever they wanted it, and they wanted it often, for it is good eating. The goats shrubbed this land completely and native grasses and white clover volunteered and covered it with a strong crop.

For the past nine years I have pastured approximately seven months each year between seventy and eighty goats, sheep, hogs, cows, and horses on this ground. If I was to base the pasture's profit at the usual rental price of $1 per head per month, it would sound like too much of a good thing to be a fact, but putting the price at 50 cents per head per month, this pasture has yielded during this nine-year period over $2,400. Thus the goats and pasture together have yielded a profit of $2,800 in nine years, or something over $300 per year from a piece of ground that was totally unproductive and valueless and would have continued so but for the use of goats.

The thought just here is that on almost every farm, especially in the hill and timber districts, there are tracts of land similar to this, waiting for the goat to turn them from eyesores and valueless investments into profitable pastures.

Angora Goats, The Land Clearers

DOG MOST DANGEROUS AND WORTHLESS OF ALL DOMESTIC ANIMALS

How 50 Head of Sheep worth $1,000 were Destroyed in One Night by Two Dogs

Henry K. Reed of Beaver County, Pennsylvania, had 54 breeding ewes worth $20 each. The night of February 8, 1917, two dogs killed 50 of them and crippled the others.

Fifty ewes, producers of food and clothing, the product of years of skill and endeavor in breeding, the pride of a good farmer, and the source of his income, totally destroyed in one night by two worthless curs.

(Photo from National Stockman and Farmer.)

One Night's Work Of Two Dogs

This story is but an incident in the history of the ravages of dogs. Farmers in all parts of the United States have found in their pastures dead, crippled, torn, bleeding, and frightened sheep, the result of raids by dogs. Sheep confined for protection in yards and houses have been raided and killed. Flocks have been destroyed outright. Flocks have been chased and worried until worthless.

Men have been financially ruined, become discouraged and

their farms made destitute—sheep husbandry has been driven out of long-settled communities and kept out of new ones because there is no protection against prowling, vicious, wolf-dogs that rant, murder and go law-free.

How long will a civilized country permit this?

DOGS THE REASON FOR SHEEP SCARCITY

In a review of 5,000 farmers in all parts of the United States, all but eighteen gave "dogs" as the main reason for the scarcity of sheep.

The dog is a carrier of hog cholera, stomach and tape worms, lice, ticks, fleas, rabies, and foot and mouth disease.

He brings contagious diseases home to the family.

He runs at large, practically unrestrained, enjoying undisputed rights, and all for what?

What has the dog family ever done that they should have more rights than their owners?

All states have "dog laws" that if enforced would lessen dog troubles, yet dog laws are nearer dead letters than anything on the statute books.

What is the remedy?

First—Enforce the laws already passed until better ones are enacted.

Then—Unqualifiedly put the dog on the same legal status as sheep, hogs, horses, and cattle. Take away the right to run at large day and night.

Compel every dog owner to keep his dog on his own premises or under his control when away from home.

A dog to be of any value to his owner must be about his owner's business.

How long would law or public opinion allow a 100-pound shote or a three-months old calf to follow its owner to town, down the street, into the store, blacksmith shop, post office, and on the way home gallop through the neighbors fields, yards, sheep folds, and feed lots?

Any one has a right to own a dog but no one has a moral right to maintain a nuisance.

Dog-proof fences are all right but let the dog owner build them. A fence that will keep dogs out will keep dogs in.

Dog chains and muzzles are inexpensive.

The use of both can be prescribed by law.

The dog problem is not solved by license.

Licensing a dog, requiring the owner to buy a collar for him, or to pay a heavy tax on him, does not keep the dog from killing sheep or being a worthless cur.

Require the dog owner to be responsible for the whereabouts of his dog.

Back this requirement with public sentiment and officers with backbone not of gristle, and sheep will come to their own.

A medium sized cow bell on every tenth sheep will help to frighten dogs away and alarm the owner and neighbors. Don't use little dinky sheep bells. They don't make enough noise. USE COW BELLS.

DOGS OR SHEEP—WHICH?
By H. S. Mobley

At the time when the woods and plains were inhabited by wild animals, some useful as a source of food and clothing, and others a menace to his safety, man found the dog a most useful and dependable aid in the chase or as a sentinel and defender. Thus originated a racial attachment that continues to the present.

But these old time conditions are changed. Man now has recourse to breeding and raising the domesticated animals, the cow, the horse, the hog, the goat, and the sheep for a large part of his meat, and clothing supply. Regarding this combination the dog has reversed his former position until now so far as sheep and goats are concerned, he has become a menace and disadvantage where he was formerly a most efficient helper. His value now is almost wholly sentimental on account of his past service.

The question to be decided now is, does his past services as a helper, give him sufficient value to justify his ravages of the flocks of sheep and goats which now are so necessary to man as a source of food and clothing?

The irresistible conclusion is that since the sheep and the goat are producers of food and clothing, and the dog produces nothing, but is a consumer of and a destroyer of one of the most important food resources, man must deny his sentimental appreciation of the dog in favor of his more pressing need of meat and clothing and substitute the sheep and the goat for the dog.

CO-OPERATIVE LAMB SELLING

Furnishes a Market for the Man With a Half Dozen Lambs. Co-Operative Selling is Successful if Well Managed

In communities where farmers raise small bunches of fat lambs they can be successfully marketed in a co-operative way by making up car lots and shipping to good markets, instead of selling to a local buyer at a sacrifice as is sometimes done.

Occasionally the local market is the best and more can be realized from a bunch of lambs by selling them a few at a time as local customers or butchers want them. When there is no good local market the co-operative selling association will solve the problem.

Such associations are successfully operating in several counties in the United States. They need not be for selling sheep or lambs alone. Lamb selling can be carried on by an association formed for the purchase or sale of any product.

Selling clubs or associations must be managed in a business-like way and the members must not be "Kickers."

Selling organizations can never be profitably operated unless some one who is competent is in charge and the members agree to abide by the decisions of the manager or directors.

The plan for selling lambs, wool, or any farm product must be worked out to fit the conditions of the locality.

In some sections where lambs are sold co-operatively the Secretary or Manager of the organization after consulting with the farmers advises either in the local paper or by letter that on a certain day a shipment of lambs will be made and for farmers to communicate with him at once as to how many and what kind of lambs they can furnish.

This gives the manager a chance to arrange for shipment and also gives the farmers a chance to sell the lambs that are then ready for market.

The smaller lambs and the ones not yet fat enough can be held until another shipment is made.

This plan of selling opens a market for the man with two or three lambs as well as for the man with twenty.

The sellers must agree to abide by the market prices for the grade to which their lambs belong.

In selling car loads a better price can usually be obtained by selling all the lambs of one grade together, regardless of who the consignor is.

Arrangement can be made to weigh each man's lambs separately after the sale, so each will get his share of the returns.

The County Agent can do a great work in helping to start a lamb selling club.

BOYS' AND GIRLS' SHEEP CLUBS

Why not start a sheep club in your community? There are pig clubs, calf clubs, poultry clubs, garden clubs and canning clubs.

Sheep clubs are an opportunity for bankers, business men, county agents, fathers and mothers, and all to encourage the boys and girls and at the same time create interest in an industry that is much needed—the raising of sheep.

The Banker can do no greater good for a lad than to loan him money to buy a pair of good ewes, and help the boy get in touch with the county agent to receive advice as to feeding and caring for his sheep.

Boys' and girls' sheep clubs will do much to create sentiment that is needed against prowling dogs.

A community with twenty-five boys and girls with a pair of good ewes each, will have to keep its dogs where they belong.

Boys who early in life learn to successfully care for sheep will develop characteristics that will be valuable, no difference what they may pursue later in life. The best sheep farmers today are the men who learned to care for sheep when they were boys. Start Sheep Clubs!

SUMMARY

The future wool and mutton supply of the United States must come from farms where small flocks of ewes are kept to grow wool and raise lambs.

The range flocks of the west are being divided and sold as homesteaders take up the pasture lands.

Sheep are profitable. The average cost of feeding a ewe for a year and her lamb until it was sold was $4.69. The income from lamb and wool was $11.15. These figures are the average of 1,000 reports from Corn Belt farmers in 1916.

Nearly every farm can keep a small flock of ewes half of the year on weeds and on grass and feed that otherwise would go to waste. Sheep improve the farm's appearance; convert waste into profit.

Sheep will eat weeds, and weed seeds eaten by sheep never grow.

The roof is the main part of a sheep house. Many farmers shelter their sheep in straw sheds.

Sheep must have dry yards and pastures. Muddy yards make sore feet. Blue vitriol dissolved in water is a good treatment for sore feet.

Don't rush into sheep raising by buying a large flock of high priced ewes.

The best way to start with sheep is to buy a few good young ewes. Get them uniform and of same breeding. Their wool and lambs will be worth more.

Sometimes the beginner can buy a few old ewes cheaply and by giving them good care get a start of lambs before the ewes get past their usefulness.

The owner of a large flock will sometimes sell the undersized lambs cheaply at weaning time. The beginner can secure a few small ewe lambs and with good care grow them into a bunch of good ewes.

At the Missouri Experiment Station lambs sired by a pure-bred ram sold for $2.85 per hundred more than lambs sired by a scrub ram. The mothers of the lambs were alike and all had the same care. This emphasizes the importance of pure-bred sires.

Ewes and lambs should have a change of pasture every two weeks to prevent stomach worms (see Page 48 for treatment). Ewes should be fed like dairy cows when they are raising lambs.

Every farm should have a patch of rape for the lambs.

Care should be used in getting the sheep accustomed to rape pasture to prevent bloating. Pasture a short time each day for several days, and at first keep sheep off after a rain or when dew is on.

There is also danger from bloat when clover or alfalfa is pastured.

There is no place where lambs will thrive better and with more profit than in the standing corn just as it begins to ripen. They will eat the lower blades that would otherwise go to waste.

Silage and alfalfa make an ideal sheep feed.

The breed depends on locality and the preference of the owner. The large mutton breeds thrive better in small flocks and they must be liberally fed. The fine wooled sheep are the hardiest and will live in large herds.

If the sheep have ticks on them they should be dipped twice—about two weeks between dippings.

Many good shepherds dip their sheep once a year whether they have ticks or not.

Sheep men are too careless about preparing wool for market. Sheep should not be permitted to get burs, chaff or straw in their wool.

Manure and filth should never be tied up in the fleeces and the tying should be done with "paper twine."

Never tie wool with binder twine.

Machine shears are best for shearing. They should be run by gas engine or farm power. Turning by hand is not satisfactory.

Dogs are a great menace to the sheep industry. Every state should have laws unqualifiedly putting the dog on the same legal status as horses, cattle, and hogs.

Where there are co-operative associations or farm bureaus, ewes can be purchased in car lots by the organization for distribution among the farms of the community. Where there is no organization farmers can join in the purchase of a car load of ewes if there are no sheep in the neighborhood.

Here is a chance for the County Agent to do good work.

A PARTIAL LIST OF SHEEP LITERATURE

BOOKS

Sheep Farming in America, by Joseph E. Wing. Published by Breeders' Gazette, Chicago.

Sheep Management, Breeds and Judging, by Frank Kleinheinz. Published by the author, Madison, Wis.

"Wool," by Edw. W. France. Published by Museum of School and Industrial Art, Philadelphia, Pa.

Modern Sheep, by Chas. E. Stewart. Published by American Sheep Breeder, Chicago.

EXPERIMENT STATION BULLETINS

Sheep Raising, Extension Circular No. 49, Pennsylvania State College, State College, Pa.

Growing and Marketing Wool, Circular No. 161, Illinois Agricultural Experiment Station, Urbana.

Docking and Castrating Lambs, Circular No. 61, Missouri Experiment Station, Columbia.

Corn Silage for Winter Feeding of Ewes and Young Lambs, Bulletin No. 147, Purdue Agricultural Experiment Station, Lafayette, Ind.

Series of Bulletins on Fattening Western Lambs, by Purdue Agricultural Experiment Station, Lafayette, Ind.

Advantage from Use of Pure Bred Ram, Circular No. 65, Missouri Experiment Station, Columbia.

Rations for Breeding Ewes, Bulletin No. 120, Missouri Experiment Station, Columbia.

The Maintenance of Breeding Ewes of Mutton and Wool Sheep, Bulletin No. 144, Pennsylvania State College. State College, Pa.

Market Classes and Grades of Sheep, Bulletin No. 129, Illinois Agricultural Experiment Station, Urbana.

FARMERS BULLETINS

Sheep Feeding, Farmers' Bulletin No. 49, U. S. Dept. Agriculture, Washington, D. C.

Raising Sheep for Mutton, Farmers' Bulletin No. 96, U. S. Dept. of Agriculture, Washington, D. C.

Replanning a Farm for Profit, Farmers' Bulletin No. 370, Page 17, "A Sheep Farm," U. S. Dept. of Agriculture, Washington, D. C.

Mutton and Its Value in the Diet, Farmers' Bulletin No. 526, U. S. Dept. of Agriculture, Washington, D. C.

The Angora Goat, Farmers' Bulletin No. 573, U. S. Dept. of Agriculture, Washington, D. C.

Breeds of Sheep for the Farm, Farmers' Bulletin No. 576, U. S. Dept. of Agriculture, Washington, D. C.

Sheep Killing Dogs, Farmers' Bulletin No. 652, U. S. Dept. of Agriculture, Washington, D. C.

Sheep Scab, Farmers' Bulletin No. 713, U. S. Dept. of Agriculture, Washington, D. C.

Sheep Raising for Beginners, Farmers' Bulletin No. 840, U. S. Dept. of Agriculture, Washington, D. C.

www.ingramcontent.com/pod-product-compliance
Lightning Source LLC
Chambersburg PA
CBHW081737220526
45468CB00008B/2138